Size

Size

How It Explains the World

VACLAV SMIL

wm

WILLIAM MORROW

An Imprint of HarperCollins*Publishers*

HarperCollins books may be purchased for educational, business, or
sales promotional use. For information, please email the
Special Markets Department at SPsales@harpercollins.com.

Originally published in Great Britain in 2023 by Penguin Random House UK.

FIRST U.S. EDITION

Library of Congress Cataloging-in-Publication Data has been applied for.

ISBN 978-0-06-332409-1

23 24 25 26 27 LBC 5 4 3 2 1

Contents

Preface

Writing a book on a big topic—be it energy, the economy, or the environment, or the history of painting, population growth, or pandemics—is always a ruthless exercise in exclusion.

The challenge (and solution!) was best summarized by Jorge Luis Borges in "On Exactitude in Science," a short story published in *Los Anales de Buenos Aires* in 1946, about cartographers in an unnamed empire who want to achieve perfection. They start with the quest for a perfect map of a single province; the finished map occupies an entire city. Their map of the empire is the size of an entire province. But that isn't good enough, and so their ultimate creation—a map of the empire that plots every point perfectly—is the size of the empire itself. Inevitably, "succeeding generations understood that this extended map was useless, and without compassion, they abandoned it to the inclemencies of the sun and of the winters."

When you write a book on a big topic, you have to deal with an unceasing flood of information. Printed descriptions and analyses accumulated by generations of traditional academic research have been augmented by new electronic sources, databases, maps, and image collections, most of them still growing rapidly. Even dedicated and protracted submersion in these oceans of information cannot assure familiarity with everything that might be considered important, and converting this inevitably incomplete understanding into a single-volume book of a reasonable length requires further omissions still. This common problem becomes even more obvious when writing not simply about big topics but about such overarching subjects as growth, scale, and size.

This book is the very opposite of those foolish attempts in Borges's story: it tries to fit so much into so few pages, but this, at least, should make it more useful than the perfect map! Even before I began this book about size I had to make—in order to keep the volume within a reasonable length—some basic compromises about its scope and depth. This is not a preemptive excuse, just an acknowledgment of reality.

Musical analogy may be helpful. The book starts *andante*, and for most of its length it continues that way. Here and there I offer some unusual, even surprising, chords, but always in a harmonious way, advancing the main theme. There are no stunning irruptions. The first chapter offers insights into the role of size in nature and in human affairs, followed by comments on the tension between small and large, on the widespread human preference for larger sizes, and on extremes of size. The second chapter focuses on our perception of size, on what we see and what we think we see (illusions are surprisingly easy to create), with a closer look at the consequences of human body height, the size that affects (to a surprisingly high degree) many aspects of our lives. The third chapter looks at relations between and among sizes—stories of proportions, symmetries, and ratios—and ends with an examination of the so-called golden ratio, which popular culture would have you believe appears everywhere. We'll see about that.

The fourth chapter, on size designs, opens with an introduction to ergonomics—the science of designing for both ease of use and for safety—and it deals with one of its most consequential applications in our restless society: the design of aircraft seats. Given the unmistakable modern trend toward larger designs of so many essential artifacts (from electricity-generating stations to wind turbines; from motor vehicles to airplanes), it is revealing to look at the growth of some of these sizes over time and at some of the limits that already restrict their further expansion. Much like organisms, artifacts have their growth limits.

The fifth chapter looks at scaling: how a particular size changes as a consequence of another change taking place. You may have never thought about this, but you will already know some of the answers: if you weigh 60 kilograms, you know that people who weigh twice as much do not have head circumferences twice as large. But have you ever thought about their hearts: are they the same size as yours—and if not, how much larger must they be? The scaling of metabolism (how energy requirements depend on body mass) in humans, other mammals, and other classes of animals is particularly important. The sixth chapter is devoted to these realities.

After that, what remains to be said about size? The more statistically minded readers might point out that, so far, I have not said anything about size distributions. You know that the Earth's features (whether mountain heights or lake areas) and living organisms (be they micro- or macroscopic), as well as countless artifacts (from the simplest tools to the most elaborate machines) come in a variety of sizes. But how are these sizes distributed? When graphed, do they display symmetry or are they highly asymmetrical? The last two chapters will answer these questions.

As opposed to my *andante* beginning, the closing chapter will be at least *allegro* (if not *presto*), as I will try to provide a summation worthy of the third decade of the 21st century: brief, size-limited entries briskly moving along. Readers looking for a grand finale ending in a stunning revelation of condensed wisdom will be greatly disappointed: there is no apotheosis here, no triumphal chords. If there is such a finale to cap a book devoted to the unruly and complex realities of size—to its measurements, perceptions, properties, changes, scaling, and distributions—it will have to wait for the next (bolder and more perceptive) interdisciplinarian to compose it, I have done my best.

I SIZE AS THE MEASURE OF ALL THINGS

Protagoras, the first (and perhaps the greatest) of Greek sophist philosophers, opened his book on truth with the famous phrase "man is the measure of all things" (πάντων χρημάτων μέτρον ἐστὶν ἄνθρωποσ; *pánton chrimáton métron estín ánthropos*). This led to a stream of arguments, explanations, and denials that shows no sign of abating.[1] I have neither the qualifications nor any reckless desire to join these philosophical debates; I simply want to stress the obvious. Philosophers deconstructing Protagoras's opening line focus on his choice of "man" (*ánthropos*)—but is not the choice of "measure" (*métron*) just as central to his pronouncement? *Ánthropos* is not an observer or a subject, nor a witness or a judge—but is the measure of all things. And, inevitably, measuring—be it explicit and using actual physical units, or implicit by comparing someone or something to a standard or to a mental image—is reckoning with size.

The perception of size is inextricably bound up with the perception of shape: the spatial form, the outward appearance, the identifiable features of an object. John Locke, in his *Essay Concerning Human Understanding*, went as far as claiming that in organized bodies (plants and animals): "The shape is that which to us is the leading quality, and most characteristical part, that determines the species . . . For however some men seem to prize their definition of animal rationale, yet should there a creature be found that had language and reason, but partaked not of the usual shape of a man, I believe it would hardly pass for a man, how much soever it were animal rationale."[2]

Nowhere is this conclusion, this inescapable condemnation by shape, better illustrated than when Jonathan Swift sent Lemuel Gulliver on his last imaginative journey to the land of exalted Houyhnhnms (intelligent horses) and repulsive human-shaped Yahoos. Gulliver could speak and reason with those wise and benevolent horses, but in their eyes his shape could elevate him no higher than being a "gentle Yahoo."[3]

Intriguingly, languages have two different ways of conceptualizing size. In ancient Greek, the language of definitions and explanations, the word for size is μέγεθος (*mégethos*: extent, largeness, bigness) and, much like the Latin *magnitudo*, it directs our minds toward the upper end of the size spectrum ("large" in Greek and Latin: *mégas*, *magnus*). Unlike the ancient Greek, Latin also has a neutral *dimensio* (dimension), but in modern European languages this division does not correspond to linguistic groups (Romance, Germanic, Slavic). Much like the English "size," the three largest Slavic languages—Russian, Ukrainian, and Polish—put it in neutral terms (*razmer*, *razmir*, and *rozmiar*: measure), but Czech and Slovenian are biased toward large sizes (*velikost* in both of them), and so are Croatian (*veličina*), the three leading Germanic languages—German (*die Grösse*), Swedish (*storlek*), and Dutch (*grootte*)—and Italian (*grandezza*).

But Italians can also say *taglia*, from the Latin *talea* (cutting) that gave us the French *taille* (originally a long and slender shape). The Japanese have three options. The newest one is among the tens of thousands of words borrowed from English and adapted to Japanese syllabic writing (*saizu* or サイズ). The first of the two classic choices, with the kanji characters borrowed from China, is the neutral 寸法 or *sunpō* (dimension, measurement). But the best of all is the evocative ancient Chinese choice of *dàxiǎo* (*daishū* in Japanese), a combination of the two simple three-stroke characters for "large" and "small": 大小.

Multilingual readers will be aware of many of these deeply embedded differences concerning the perception of size, but, while interesting, it does not matter if the concept is neutral or not. In fundamental physical terms, size is the only value needed to specify scalar quantities—those that are defined (unlike force or velocity) without any direction: length (width, height, circumference), area, volume, mass, energy. Once we have accurate tools for measuring things (for example, measuring tapes for body height; laser distance sensors in construction), and once we agree on the required level of accuracy (body height measured to the nearest centimeter) then sizes can be ascertained routinely, repeatedly, and reliably.[4]

Of course, there are some physical variables where the answers will vary a great deal depending on the measuring standard, a problem noted originally by Lewis Fry Richardson in the early 1950s and elaborated by Benoit Mandelbrot in his often-cited paper published in *Science* in 1967, "How long is the coast of Britain?"[5] Unlike the measurement of a linear distance between two fixed points, this question has no definite answer; the results depend on the scale of the map used. Obviously, a map at 1:10,000,000 (with every centimeter corresponding to 100 kilometers) will display a lot less detail, and will result in significantly shorter coastline length, than when using a hiking map where every centimeter stands for just 1 kilometer (a scale of 1:100,000).

And the results would be much higher still if we were to try to survey, by walking in short steps (or by putting heel to big toe), every meandering irregularity of the actual coastline; and even more so if we should attempt the task of measuring it in 1-millimeter increments— that, being the size of small sand grains and soil particles, would be practically impossible. On larger scales, Google Earth has made these exercises easy. All you have to do is to click "+" in the lower right corner of the display to see the imagined area in progressively greater detail, all the way from looking at the planet from an altitude of 22,252 kilometers to distinguishing (in some instances) features smaller than 1 meter. Clearly, the concept of definite size (a single immutable value) is not applicable when looking at physical features of the Earth: the length of a coastline, or of an international border, is not a fixed quantity; it has a fractal dimension, a ratio expressing how details change with the scale used to measure them.[6] But in the real world there are many more important challenges in measuring size than the seemingly indeterminate length of a coastline.

Perhaps most notably, measuring the sizes of bodies or of important social and economic variables like gross domestic product and income (information needed to judge progress on the individual level and to make rational policy decisions) poses many problems of accuracy and compatibility.[7] In a completely salaried western urban society, the problem of calculating a country's annual GDP or income might seem to be as simple as adding up all economic activities or all salaries. But even in the richest countries, this leaves out a great deal of earnings from the underground (shadow) economy, from backhanded cash-only payments, and from the lucrative drug and sex trades, for example.[8] Looking elsewhere, we should set down some rules for how to deal with subsistence farmers and bartering families in Africa and Asia.[9] And when that is done, how do we compare these national monies: simply by using official, and often heavily manipulated, exchange rates, or by resorting to complex conversions to purchasing power parities?[10]

No matter if perfectly measured or poorly defined, size matters in a myriad of quotidian ways. Everyday life depends on using and encountering appropriate, often carefully standardized, sizes of clothes, utensils, tools, machines, and structural components. We rely on expected averages (or minimal dimensions); we replicate them, conform to them, and expect to encounter them again and again. We are reminded of these unconscious expectations only when matters shift out of the expected size range: having ill-fitting clothes on a transatlantic flight; struggling with a non-ergonomic tool that is difficult to wield; encountering a non-standard step rise when moving heavy furniture. Whether we like it or not, we move through a world tightly defined by size.

Modern societies have tried to limit these problems by standardizing sizes. While most people may not know the exact measurements, they are aware of many standards from their everyday experience, and people who travel or move abroad notice differences between countries.[11] Stairs in homes and public places are rarely tricky to negotiate because the risers (the vertical bit) and treads (the bit you stand on) have to conform to national or international standards. For example, the International Building Code requires risers no higher than 178 millimeters (7 inches); and, similarly, the US Occupational Safety and Health Administration prescribes a minimum of 6 and a maximum of 7.5 inches (152.4–190.5 millimeters).[12] America's most popular construction lumber, a piece of two-by-four, actually does not measure 2 × 4 inches. It used to, but now the additional milling and planing, to give the lumber a better appearance, reduces it to 1.5 × 3.5 inches (38 × 89 millimeters).[13] And Americans notice immediately that the standard European paper size for writing business letters (A4) is slightly different: a bit narrower and a bit longer (210 × 297 millimeters vs. 216 × 280 millimeters).[14]

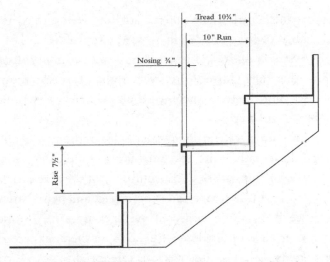

You would quickly notice if risers and treads came in random dimensions.
The drawing shows the standard US specifications given in inches and their fractions.

In our modern and overwhelmingly urban societies, with their
high population densities and extensive transportation networks, we
now specify scores of measurements in order to provide basic comfort
and safety. Mass-built dwellings must have minimum room heights—
in the US it is 8 feet, the height of a dry wall sheet—and many countries
have long prescribed a minimum floor area per person, which differs
even among affluent countries. For example, most people do not realize that France remained relatively poor until the 1950s: the norm that
applied to French low-rent dwellings (*habitations à loyer modéré*) specified a minimum of 35 square meters for a two-room apartment both
in 1922 and in the early 1950s, and no more than 45 square meters half
a century later—and in Japan the minimum requirement remains at a
mere 25 square meters per person, compared to the nearly 70 square
meters of an average newly built detached house in America (211
square meters in 2021, divided by an average family size of 3.13).[15]

We become easily habituated to these standards and prevalent sizes
and expect their repeated appearance, but, if needed, we learn to
cope with new surroundings conforming to different rules. But we
always remain alert to departures from expected sizes. This, as I will
explain in detail in the seventh chapter, is due to the fact that the sizes

of living organisms and their specific attributes—be they heights of newborn babies, wingspans of albatrosses, circumferences of sequoia trunks, or weights of Italian pensioners—are normally distributed. This means that they cluster (often very tightly) around their means, and when we plot the frequencies of their occurrence, we get a symmetrical bell-shaped curve.

Most people are not formally aware of this ubiquitous reality—and very few could actually write down the fairly complex equation specifying the curve of this statistical distribution. But they have internalized it subconsciously through repeated experiences, and hence any substantial departure from those expected averages is readily noticed and instantly processed as unusual: we'll always spot the unexpectedly large or surprisingly small. This unconscious categorizing goes further. We might assign extremes of size to categories like miniatures (tiny, minuscule, microscopic) or to the class of giants (enormities, monsters). Obviously, this can work only where we're aware of the norms. Businessmen in Manhattan, acutely aware of body heights (as I will explain in the next chapter, height is a clear marker of corporate "leadership"), would class—based on their experience with omnipresent pigeons, and with Central Park's ducks and geese—a Laysan albatross among large-wingspan birds (they average about 2 meters). But only a birder among them would know that there are albatrosses with even larger wingspans (the wandering albatross, up to 3.5 meters).[16]

Between large and small

We are a highly social species and, in the most basic survival terms, it helps us to fit in and to cope better if we are at least of average height and if the size of our intellect is not in the lowest quartile of normal distribution. Larger size is a critical prerequisite of complexity: small and isolated prehistoric groups of foragers (where everybody's one role beyond infancy was to secure enough food) could never develop social complexity, with its elaborate stratification and thousands of job descriptions, and reach the technical prowess of large urban societies.

Large size, be it of ungulate herds in Africa or modern manufacturing and electronic media companies, conveys manifold competitive advantages: both Serengeti wildebeest and Google are doing well. At about 1.5 million individuals, the Serengeti's enormous wildebeest herd is the world's largest group of massive herbivores to perform a lengthy annual migration—and Google's share of the search engine market (about 93 percent worldwide) is responsible for the company's high market capitalization: in early 2022 it was close to $2 trillion, larger than the annual GDP of Russia or Brazil.[17]

We spend our lives mentally toggling between small and large, but the two categories evoke different reactions. For all but a small fraction of our evolution, only natural phenomena—storms, fires, floods, earthquakes—brought experiences on a large, often overwhelming and deadly, scale.[18] As long as we were living in small groups foraging and scavenging for food, and as long as our tools were limited to the simplest stone implements, we could not leave any large-scale structures or durable objects behind. But given the very similar brains we share with these prehistoric hunters and gatherers who roamed Africa and Eurasia tens of thousands of years ago, we can be sure that our species imagined creatures and events on a large and frightening scale, and brought them to life in tales that were passed down for generations before the first fictional narratives were inscribed in cuneiform characters into Mesopotamian clays nearly 5,500 years ago.[19]

This continuum of imaginative fiction runs from the earliest recorded tales to modern storytelling, from the monsters of ancient legends to the box-office records generated by the computer-animated battles of Godzilla and King Kong.[20] Encounters with terrifying monsters are featured in the two oldest preserved stories of the human imagination: the travails of the Sumerian king Gilgamesh, and the adventures of Odysseus and his companions returning from Troy. On his quest, Gilgamesh must face Humbaba, a monstrous giant whom Enlil, the chief Sumerian god, assigned to protect the Cedar Forest and to terrorize human beings: "When he roars it is like the torrent of the storm, his breath is like fire, and his jaws are death itself." And after Gilgamesh slays Humbaba he encounters another monstrous creature:

a "sombre-faced man-bird . . . his was a vampire face, his foot was a lion's foot, his hand was an eagle's talon."[21]

Odysseus, on his long homeward journey, has to outsmart another giant—one-eyed Polyphemus. After the giant's single eye is blinded by a burning stake and Odysseus taunts him from the escaping ship, Polyphemus "ripped off the peak of a towering crag, he hurled it so hard that the boulder landed just in front of our dark prow . . . a tidal wave from the open sea."[22] What a frightening, out-of-scale stimulus for our imagination. The Bible's most famous giant is Goliath, slain by David's slingshot-hurled stone. His existence could be explained as a familial genetic affliction.[23]

Monstrous-sized creatures (giants and dragons who were often Hydra-like, multiheaded) with supernatural capacities—from displays of enormous power to fire breathing—have always inhabited legends and folk tales. When the Brothers Grimm (Wilhelm and Jacob) collected their German versions and, in 1812 and 1815, published them in the classic two-volume compendium *Kinder- und Hausmärchen*, their stories echoed with giants' footsteps but also had some diminutive creatures.[24] Their giants sit on top of mountains, roast sheep and oxen, rob, murder, and ravage people, but others behave timidly, even carrying men across enormous distances; they suckle thumb-size creatures so that they, too, can become giants; they are killed by brave knights. In their story "The Fisherman and His Wife" we are even given precise measurements of fictional extremes of size: the guards stand in two rows, "each being smaller than the one before him, from the biggest giant, who was two miles high, to the very smallest dwarf, just as big as my little finger."

Besides assorted folk tales, perhaps the two most famous cases of English writings about extreme sizes and imaginary size transformations are the aforementioned Jonathan Swift's *Gulliver's Travels* and Lewis Carroll's *Alice's Adventures in Wonderland*.[25] I will turn to Gulliver's experiences among tiny Lilliputians and giant Brobdingnagians in the fifth and sixth chapters of this book, when we look more closely at the scaling of bodies and their metabolisms. I will explain what Swift got right, and point out some of the key size assumptions and calculations that he got wrong as he tried to present consistent images of miniaturized and gigantic people.

While Swift's classic can serve as a revealing and highly entertaining introduction to the realities of size and the complexities of metabolic scaling, the metamorphoses of Carroll's Alice are simply ways to advance a story, but they offer an entertaining entry into the all-too-real world of visual illusions. Alice enters the "large" rabbit hole as herself, but she must, somehow, shrink immediately upon entering it (even large rabbit holes are not large enough to accommodate small girls). She manages this, falls, eventually lands, and follows a nattily dressed White Rabbit. Then she starts a series of deliberate size changes, first by drinking from a marked bottle to shrink in order to enter a magic garden through a low door, but she forgets to pick up the key on the table she can no longer reach and has to eat a tiny cake with "EAT ME" spelled in currants—this sees her body telescopically expanding and her head striking the room's roof. She later tries to make herself the right size by following the Caterpillar's impossible advice to eat a mushroom of which one side will make her grow taller and the other side will make her grow shorter, which leads to her first shrinking smaller still and then growing uncontrollably.

Remarkably, reality imitates art, and medical literature describes incidents of bizarre changes in size perception known as Alice in Wonderland syndrome. This illusion of large body size was first mentioned briefly by Hermann Oppenheim in 1913, described in detail by Caro Lippman in 1952, and named by British psychiatrist John Todd in 1955.[26] One of Lippman's patients told him that she got "all tired out from pulling my head down from the ceiling. My head feels like a balloon; my neck stretches and my head goes to the ceiling. I've been pulling it down all night long." Besides altered body height, the sufferers also see illusionary shapes (metamorphopsia) and perceive elongations or shrinkage of body parts or changing dimensions and motions of objects. These hallucinations take place, infrequently, during some migraine headaches, partial seizures, infections, and intoxications.

In the real world, the attraction of going big was evident even when our capabilities were limited to human muscles and to such simple machines as levers and ramps (inclined planes made it easier to elevate heavy objects). We deployed these mechanical aids with great ingenuity, to build on scales as large as the restrictions posed by our bodies and muscle-powered tools and machines allowed us to do. During prehistory, our universal tendency to go beyond the human scale was channeled into some impressive stone monuments, with large blocks quarried and often brought from afar to be erected as monumental structures. Stonehenge's largest monolith (about 45 centuries old) weighs about 30 tons—smaller bluestones of around 2 tons were brought from Wales's Preseli Hills, some 220 kilometers away. Brittany's equivalent, the Grand Menhir Brisé, is much older—erected about 6,700 years ago—and weighs 340 tons.[27]

We can only guess at the logistics of such enterprises; history offers many more of these admirable accomplishments. Tellingly, all seven wonders of the ancient world made the list due to their unusually large size, and the oldest one of them (Khufu's pyramid at Giza) also remained the tallest structure, at 139 meters, until 1311—when builders completed the spire of Lincoln Cathedral.[28] Famous large structures in the Americas include earth and stone Mesoamerican pyramids built about 1,000 years ago; massive stone structures at Ollantaytambo and at Sacsayhuamán, both from the 15th century; and enormous geoglyphs in the Peruvian and Chilean deserts.[29]

Nearly all of the most famous, most admired, and most visited structures—be they of antiquity (the Parthenon on the Athenian Acropolis; Rome's Pantheon with its large dome, or its Colosseum with its gruesome history), the Middle Ages (tall-spired cathedrals, massive castles), the Renaissance (Florence's Cattedrale di Santa Maria del Fiore with its enormous cupola, ingeniously designed by Filippo Brunelleschi to be built without scaffolding; or Rome's San Pietro basilica), or of the 19th century (including the two most visible landmarks in Paris, the Sacré-Coeur and the Eiffel Tower)—are distinguished by their large size.[30] In comparison to crowds besieging such sites, very few admirers go out of their way to visit the small pointed pyramids built in northern Sudan by Kushite kingdoms (contemporaneous with ancient Egypt), or walk up the steep Gianicolo hill to admire one of the perfectly proportioned masterpieces of the Italian renaissance, Donato Bramante's 1502 Tempietto, a small commemorative martyrium (erected on the spot where tradition places St. Peter's crucifixion) squeezed into the courtyard of Rome's San Pietro in Montorio.[31]

Filippo Brunelleschi's dome of the Cattedrale di Santa Maria del Fiore in Florence (internal diameter 43 meters) and Donato Bramante's dome of the Tempietto in Rome (internal diameter just 4.5 meters).

Obviously, large has always been a byword for importance and grandeur in the human imagination. Large impresses us, it awes us, it frightens us; and, as far as human artifacts are concerned, it also inspires us to go further, to set new limits, to design larger structures (taller skyscrapers), larger transportation machines (be it jetliners or cruise ships), and, alas, larger political and economic empires. We set much-welcomed records (larger industrial facilities making consumer products more affordable by lowering their unit costs) as well as pursuing many dubious accomplishments (including a litany of excessively sized projects that had to be abandoned). I will examine this quest in some detail in the next section.

The preference for large sizes is evident early in life, and as grown-ups we engage in a great deal of size assessments and in larger-size seeking. A human figure is usually the first object infants will draw, and they draw it large, often filling the page. Children of all ages draw figures they like (say their doting mother or father) larger than those they don't (an annoying aunt or uncle, perhaps).[32] Enduring size assessments range from the lines penciled on a doorframe marking our growing childhood height, to comparing sizes of salaries and boasting about house areas. Peasants leave their villages for cities, and in the globalized world families emigrate to distant countries in order to multiply the size of their earnings—and, as an unintended effect, to have taller children too.

We have two studies demonstrating the remarkably rapid effect of better living conditions (nutrition, healthcare, housing). In 2005, Italian researchers found that children born to Chinese immigrant parents in Bologna were not only taller than children born and living in China, but during their first year their weight and height were above those of Italian children and became comparable afterward.[33] A similar outcome has been reported in England. Indian adults who migrated to England are 6–7 centimeters taller than the average in India. This still leaves them shorter than the English mean, but this difference disappears among their younger sons and daughters who, at age 2–4 years, are 6–8 percent taller than their counterparts in India. This is a remarkably rapid catch-up, even more so given the fact that the children of parents who came from India have lower (about 400 grams on average) birthweights.[34]

And, switching from the personal to the corporate, "large," "larger," and preferably "the largest" have become the most desirable adjectives to describe trajectories to success. Some providers of limited-edition luxury items aside, no company has become a global leader by drastically capping its output and aspiring to remain modest in size.[35] And there is nothing new about this trend toward larger sizes: the evolution of living organisms has supplied many precedents. What's new is the ubiquity and the pace of the modern quest for larger sizes. This accelerated trend began during the latter half of the 19th century, powered by industrialization, and its intensification through the 20th century created our modern world of record-breaking sizes and superlatives.

Before taking a closer look at the modern quest for larger size, I must make a few basic observations about small size. Obviously, small size carries its own connotations and reactions, and the skills and the perseverance of dedicated artists and craftsmen in the past—and even more so the astonishing advances of modern science have been able to produce artifacts on ever-smaller scales which have elicited a great deal of admiration. Here is an offhand selection of such items: Chinese puzzle balls (spheres inside a sphere) that seem quite impossible to carve from a single piece of ivory; wooden models of ships that get painstakingly assembled inside bottles; entire scaled-down cityscapes that are accessible to tourists (the most famous one, Madurodam in The Hague, is a 1:25 model of famous Dutch sights); the offerings of Fake Food Japan, with its scores of small-size replicas of Japanese dishes and beverages, which many collectors delight in; the ever-smaller components electronic engineers crowd onto silicon chips, enabling the miniaturization of electronic devices from mobile phones to tracking devices small enough to be attached to the backs of small insects.[36]

But accomplishments on miniature scales do not evoke the same emotional response as those on the human scale and those that far surpass it. A book-size still life of flowers does not have the same emotional impact as large-size portraits of individuals or groups, as depictions of mythical scenes or historical events, or as landscapes depicting views akin to those we experience outside. Millions of people do not annually visit the Louvre or Prado to see their large collections of miniatures and jewelry, but to be captivated by life-size portraits by Diego Velázquez and Francisco Goya.[37] The *Guinness Book of World Records* lists all kinds of useful and bizarre facts, but in 2021 only one among its website's 18 most frequently visited records referred to something excessively small (the world's shortest woman). The rest were all maxima: longest, highest, largest, most, and also fastest, widest, and oldest.[38]

Miniatures can impress and entertain—but they cannot awe us: that powerful sentiment is reserved for large sizes. How can miniatures compare with the sight of a dark cumulonimbus cloud that has grown for hours in height on a hot summer afternoon and that is just on the point of turning into a powerful tornado?[39] What is a tiny carving compared to seeing a massive tugboat alongside an enormous container ship in Los Angeles harbor, loaded ten stories above its deck with thousands of steel cases, having just crossed the Pacific Ocean?[40]

"Going small" evokes emotions that are very different from those elicited by "going large"—and, in any case, for an unaided eye such experiences are soon cut off by the limits of human vision. Look at your hand, the part of your body that is most often in your field of vision. For an average adult, the width of a pinky is about 1 centimeter or 10 millimeters, and the best naked eye can distinguish hair-like widths down to about 0.04 millimeters.[41] The quotient of these widths (0.04/10) is 0.004, which means that you can discern nothing that is less than 1/250 (0.4 percent) as wide as your pinky. In contrast, when going large, we can see a small raindrop (1 millimeter), rainwater falling over a street curb (height of 10 centimeters), a shallow weir (raising water by a meter), a mountain stream tumbling over large boulders (10 meters tall), Angel Falls in Venezuela (just 21 meters short of 1 kilometer), and an enormous vertical cumulonimbus cloud rising 10 kilometers into the lower stratosphere. The quotient of these extremes (in millimeters) is 10,000,000/1; an imposing cumulonimbus is 10 million times larger than a small raindrop.[42]

Going small (increasing the number of components on a microprocessor) has sustained the post-1965 growth of electronics (I will deal with this trend in the fourth chapter, on size designs). That, in turn, has been a key contributor to the global economic growth of the past two generations, and also (paradoxically) a key enabler of going large. The quest for larger artifacts—screens, cars, ships, airplanes, turbines, buildings, cities—has become much easier thanks to the miniaturization of solid-state (transistors, integrated circuits, microprocessors) electronics.

Modernity's infatuation with larger sizes

A single human lifetime will have witnessed many obvious examples of this trend in sizes. Motor vehicles are the planet's most numerous heavy mobile objects. The world now has nearly 1.5 billion of them, and they have been getting larger: today's bestselling pickup trucks and SUVs are easily twice or even three times heavier than Volkswagen's *Käfer*, Fiat's *Topolino*, or Citroën's *deux chevaux*—family cars whose sales dominated the European market in the early 1950s.

Sizes of homes, refrigerators, and TVs have followed the same trend, not only because of technical advances but because the post–Second World War sizes of national GDPs, so beloved by the growth-enamored economists, have grown by historically unprecedented rates, making these items more affordable. Even when expressed in constant (inflation-adjusted) monies, US GDP has increased 10-fold since 1945; and, despite the postwar baby boom, the per capita rate has quadrupled.[43] This affluence-driven growth can be illustrated by many other examples, ranging from the heights of the highest skyscrapers to the capacity of the largest airplanes or the multistoried cruise ships, and from the size of universities to the size of sports stadiums. Is this all just an expected, inevitable replication of the general evolutionary trend toward larger size?

We know that life began small (at the microbial level as archaea and bacteria that emerged nearly 4 billion years ago), and that, eventually, evolution took a decisive turn toward larger sizes with the diversification of animals during the Cambrian period, which began more than half a billion years ago. Large size (increased body mass) offers such obvious competitive advantages as increased defense against predators (compare a meerkat with a wildebeest) and access to a wider range of digestible biomass, outweighing the equally obvious disadvantages of lower numbers of offspring, longer gestation periods (longer time to reach maturity), and higher food and water needs.[44] Large animals also live (some exceptions aside—some parrots make it past 50 years!) longer than smaller ones (compare a mouse with a cat, a dog with a chimpanzee). But at its extreme the relationship is not closely mass-bound: elephants and blue whales do not top the list; Greenland sharks (more than 250 years), bowhead whales (200 years), and Galapagos tortoises (more than 100 years) do.

The evolution of life is, indeed, the story of increasing size—from solely single-celled microbes to large reptiles and modern African megafauna (elephants, rhinos, giraffes). The maximum body length of organisms now spans the range of eight orders of magnitude, from 200 nanometers (*Mycoplasma genitalium*) to 31 meters (the blue whale, *Balaenoptera musculus*), and the extremes of biovolume for these two species range from 8×10^{-12} cubic millimeters to 1.9×10^{11} cubic millimeters, a difference of about 22 orders of magnitude.[45]

The evolutionary increase in size is obvious when comparing the oldest unicellular organisms, archaea and bacteria, with later, larger, protozoans and metazoans. But average biovolumes of most extinct and living multicellular animals have not followed a similar path toward larger body sizes. The average sizes of mollusks and echinoderms (starfish, urchins, sea cucumbers) do not show any clear evolutionary trend, but marine fish and mammals have grown in size.[46] The size of dinosaurs increased, but then diminished as the animals approached extinction. The average sizes of arthropods have shown no clear growth trend for half a billion years, but the average size of mammals has increased by about three orders of magnitude during the past 150 million years.

Analyses of living mammalian species show that subsequent generations tend to be larger than their parents, but a single growth step is inevitably fairly limited.[47] In any case, the emergence of some very large organisms has done nothing to diminish the ubiquity and importance of microbes: the biosphere is a highly symbiotic system based on the abundance and variety of microbial biomass, and it could not operate and endure without its foundation of microorganisms.[48] In view of this fundamental biospheric reality (big relying on small), is the anthropogenic tendency toward objects and design of larger sizes an aberration? Is it just a temporary departure from a long-term stagnation of growth that existed in premodern times as far as both economies and technical capabilities were concerned, or perhaps only a mistaken impression created by the disproportionate attention we pay nowadays to the pursuit and possession of large-size objects, from TV screens to skyscrapers?

The genesis of this trend is unmistakable: size enlargements have been made possible by the unprecedented deployment of energies, and by the truly gargantuan mobilization of materials. For millennia, our constraints—energies limited to human and animal muscles; wood, clay, stone, and a few metals as the only choices for tools and construction—circumscribed our quest for larger-designed sizes: they determined what we could build, how we could travel, how much food we could harvest and store, and the size of individual and collective riches we could amass.[49] All of that changed, rather rapidly and concurrently, during the second half of the 19th century.

At the century's beginning, the world had very low population growth. It was still energized by biomass and muscles, supplemented by flowing water turning small wheels and wind-powering mills as well as relatively small ships. The world of 1800 was closer to the world of 1500 than it was to the mundane realities of 1900.[50] By 1900, half of the world's fuel production came from coal and oil, electricity generation was rapidly expanding, and new prime movers—steam engines, internal combustion engines, steam and water turbines, and electric motors—were creating new industries and transportation capabilities. And this new energy abundance was also deployed to raise crop yields (through fertilizers and the mechanization of field tasks), to produce old materials more affordably, and to introduce new metals and synthetics that made it possible to make lighter or more durable objects and structures.

This great transformation only intensified during the 20th century, when it had to meet the demands of a rapidly increasing population.

Despite the two world wars and the Great Depression, the world's population had never grown as rapidly as it did between 1900 and 1970.[51] Larger sizes of everything, from settlements to consumer products, were needed both to meet the growing demand for housing, food, and manufactured products and to keep the costs affordable. This quest for larger size—larger coal mines or hydro stations able to supply distant megacities with inexpensive electricity; highly automated factories producing for billions of consumers; container vessels powered by the world's largest diesel engines and carrying thousands of steel boxes between continents—has almost invariably coincided with lower unit costs, making refrigerators, cars, and mobile phones widely affordable. But it has required higher capital costs and often unprecedented design, construction, and management efforts.[52]

Too many notable size records have been repeatedly broken since the beginning of the 20th century, and the following handful of increases (all quantified by 1900–2020 multiples, calculated from the best available information) indicate the extent of these gains. Capacity of the largest hydroelectricity-generating station is now more than 600 times larger than it was in 1900.[53] The volume of blast furnaces—the structures needed to produce cast iron, modern civilization's most important metal—has grown 10 times, to 5,000 cubic meters.[54] The height of skyscrapers using steel skeletons has grown almost exactly nine times, to the Burj Khalifa's 828 meters.[55] Population of the largest city has seen an 11-fold increase, to Greater Tokyo's 37 million people.[56] The size of the world's largest economy (using the total in constant monies): still that of the US, now nearly 32 times larger.[57]

But nothing has seen a size rise comparable to the amount of information we have amassed since 1900. In 1897, when the Library of Congress moved to its new headquarters in the Thomas Jefferson Building, it was the world's largest depository of information and held about 840,000 volumes, the equivalent of perhaps no more than 1 terabyte if stored electronically.[58] By 2009 the Library had about 32 million books and printed items, but those represented only about a quarter of all physical collections, which include manuscripts, prints, photographs, maps, globes, moving images, sound recordings, and sheet music, and many assumptions must be made to translate these holdings into electronic storage equivalents: in 1997 Michael Lesk estimated the total size of the Library's holdings at "perhaps about 3 petabytes," and hence at least a 3,000-fold increase in a century.[59]

Moreover, for many new products and designs it is impossible to calculate the 20th-century increases because they only became commercialized after 1900, and subsequently grew one, two, or even three orders of magnitude. The most consequential examples in this category include passenger air-travel (Dutch KLM, the first commercial airline, was established in 1919); the preparation of a wide variety of plastics (with most of today's dominant compounds introduced during the 1930s); and, of course, advances in electronics that made modern computing, telecommunications, and process controls possible (the first vacuum-tube computers used during the Second World War; the first microprocessors in 1971).[60] While these advances have been creating very large numbers of new, small companies, increasing shares of global economic activity have been coming from ever-larger enterprises. This trend toward larger operating sizes has affected not only traditional industrial production (be it of machinery, chemicals, or foods) and new ways of automated product assembly (microchips or mobile phones), but also transportation and a wide range of services, from banks to consulting companies.[61]

This corporate aggrandization is measurable from the number and the value of mergers, acquisitions, alliances, and takeovers. There was a rise from fewer than 3,000 mergers—worth in total about $350 billion—in 1985 to a peak of more than 47,000 mergers worth nearly $5 trillion in 2007, and each of the four pre-COVID years had transactions worth more than $3 trillion.[62] Car production remains fairly diversified, with the top five (in 2021 by revenue: Volkswagen, Toyota, Daimler, Ford, General Motors) accounting for just over a third of the global market share, compared to about 80 percent for the top five mobile phone makers (Apple, Samsung, Xiaomi, Huawei, Oppo) and more than 90 percent for the Boeing–Airbus commercial jetliner duopoly.[63]

But another size-enlarging trend has been much in evidence: increases in size that have nothing to do with satisfying the needs of growing populations, but instead serve as markers of status and

conspicuous consumption. Sizes of American houses and vehicles provide two obvious, and accurately documented, examples of this trend,
and while imitating the growth of housing has been difficult in many
countries (including Japan and Belgium) for spatial and historical reasons, the rise of improbably sized vehicles has been a global trend.

A Ford Model T—the first mass-produced car, introduced in 1908
and made until 1927—is the obvious baseline for size comparisons.[64]
The 1908 Model T was a weakly powered (15 kilowatts), small (3.4
meters), and light (540 kilograms) vehicle, but some Americans born in
the mid-1920s lived long enough to see the arrival of improbably sized
and misleadingly named sports utility vehicles that have become global
favorites. The Chevrolet Suburban (265 kilowatts, 2,500 kilograms, 5.7
meters) wins on length, but Rolls Royce offers a 441-kilowatt
Cullinan and the Lexus LX 570 weighs 2,670 kilograms.[65]

These size gains boosted the vehicle-to-passenger weight ratio
(assuming a 70-kilogram adult driver) from 7.7 for the Model T to
just over 38 for the Lexus LX and to nearly as much for the Yukon
GMC.[66] For comparison, the ratio is about 18 for my Honda Civic—
and, looking at a few transportation alternatives, it is just over 6 for a
Boeing 787, no more than 5 for a modern intercity bus, and a mere
0.1 for a light 7-kilogram bicycle. Remarkably, this increase in vehicle size took place during the decades of heightened concern about
the environmental impact of driving (a typical SUV emits about 25
percent more greenhouse gases than the average sedan).

This American preference for larger vehicles soon became another
global norm, with SUVs gaining in size and expanding their market
share in Europe and Asia.[67] There is no rational defense of these
extravaganzas: larger vehicles were not necessitated either by concerns for safety (scores of small- and mid-size cars get top marks for
safety from the Insurance Institute for Highway Safety) or by the
need to cater to larger households (the average size of a US family has
been declining).[68]

And yet another countertrend involving the shrinking size of
American families has been the increasing size of American houses.
Houses in Levittown, the first post–Second World War large-scale
residential suburban development in New York, were just short of

70 square meters; the national mean reached 100 in 1950, topped 200 in 1998, and by 2015 it was a bit above 250 square meters, slightly more than twice the size of Japan's average single-family house.[69] American house size has grown 2.5 times in a single lifetime; average house mass (with air conditioning, more bathrooms, heavier finishing materials) has roughly tripled; and the average per capita habitable area has almost quadrupled. And then there are the US custom-built houses whose average area has now reached almost 500 square meters.[70]

As expected, larger houses have larger refrigerators and larger TV screens. Right after the Second World War, the average volume of US fridges was just 8 cubic feet; in 2020 the bestselling models made by GE, Maytag, Samsung, and Whirlpool had volumes of 22–25 cubic feet.[71] Television screens started as smallish rectangles with rounded edges; their dimensions were limited by the size and mass of the cathode-ray tube (CRT). The largest CRT display (Sony PVM-4300 in 1991) had a 43-inch diagonal display but it weighed 200 kilograms.[72] In contrast, today's popular 50-inch LED TV models weigh no more than 25 kilograms. But across the globe, the diagonals grew from the post–Second World War standard of 30 centimeters to nearly 60 centimeters by 1998 and to 125 centimeters by 2021, which means that the typical area of TV screens grew more than 15-fold.[73]

Undoubtedly, many larger sizes make life easier, more comfortable, and more enjoyable, but these rewards have their own limits. And there is no evidence for concluding that oversize houses, gargantuan SUVS, and commercial-size fridges have made their owners happier: surveys of US adults asked to rate their happiness or satisfaction in life actually show either no major shifts or long-term declines since the middle of the 20th century.[74] There are obvious physical limits to all of these excesses, and in the fourth chapter I will examine some important long-term growth trends to show that the sizes of many designs have been approaching their inevitable maxima as S-shaped (sigmoid) curves are reaching the final stages of their course.

This new, nearly universal, worship of larger sizes is even more remarkable given the abundance of notable instances when larger sizes are counterproductive. Here are two truly existential examples. Excessive childhood weight is highly consequential because the burden of early onset obesity is not easily shed later in life.[75] And on the question of height, armies have always had height limits for their recruits; a below-average size was often a gift, as it prevented a small man (or a very tall one!) getting drafted and killed in pointless conflicts.[76]

Large countries pose their own problems. If their territory encompasses a variety of environments, they are more likely to be able to feed themselves and have at least one kind of major mineral deposit, though more often several. This is as true of Russia (the world's largest nation) as it is of the USA, Brazil, China, and India. But nearly all large nations tend to have larger economic disparities than smaller, more homogeneous countries do, and tend to be riven by regional, religious, and ethnic differences.[77] Examples include the North-South divide in the US; Canada's perennial Quebec separatism; Russia's problems with militant Islam (the Chechen war, curiously forgotten, was one of the most brutal post–Second World War conflicts); India's regional, religious, and caste divisions. Of course, there are counterexamples of serious disparities and discord among small-size nations—Belgium, Cyprus, Sri Lanka—but those inner conflicts matter much less for the world at large than any weakening or unraveling of the largest nations.[78]

But the last 150 years have not only witnessed a period of historically unprecedented growth of sizes, but also the time when we have finally come to understand the real size of the world, and the universe, we inhabit. This quest has proceeded at both ends of the size spectrum, and by the end of the 20th century we had, finally, a fairly satisfactory understanding of the smallest (at the atomic and genomic levels) and the largest (size of the universe) scale. How did we get there?

Extremes, and how we got to know them

Energy and mass define the universe; size is its most dominant structural variable. Enormous distances separate galaxies (on average, they are about 10 million light-years apart); planetary orbits in our solar system have radii ranging from tens of millions to hundreds of millions of kilometers. The Earth's physical environment and its intricate life-webs are fundamentally determined by the planet's distance from the Sun and by its size. Surprisingly slight differences in the orbital distance (just a few percent closer or further) would put us beyond the optimal habitable zone of bearable average temperatures (and daily and nightly extremes). A much smaller planet would have insufficient gravity to hold its life-protecting atmosphere, while gravity on a much larger planet would impede animal and human locomotion.[79]

Understanding all of this took millennia of slow advances with, as in many other instances, accelerated progress during the past century and half. To begin with, during antiquity and throughout the Middle Ages, the Earth was the center of the perceived universe. This Ptolemaic model—positing the finite universe bounded by the sphere with fixed stars (with nothing beyond), and with the Sun, Moon, and five planets (Mercury, Venus, Mars, Jupiter, Saturn) orbiting the Earth with the entire universe visible—lasted since its formulation during the 4th century BCE until it was replaced by the heliocentric model, which was actually known and dismissed during antiquity and was convincingly advocated by Nicolaus Copernicus in the early 16th century.[80] But the erroneous geocentric belief did not stand in the way of providing fairly good numbers of the Earth's size and the distances between the Earth and the Moon, and the refutation of the geocentric model some 1,700 years later did not immediately improve the original Greek (and later Arab) calculations of the size of the visible universe.[81]

In the 3rd century BCE, Eratosthenes, relying on a shadow-casting vertical stick in his hometown of Alexandria, and the distance between that city and Aswan—where the Sun stood at 90 degrees during noon of the summer solstice—calculated the Earth's circumference to be 50 times the distance between the two cities. Since we do not know the precise length of a *stadion*, the unit of distance he used, we cannot pinpoint the error in his calculation; all we know is that he got the order of magnitude right. In contrast, the first known calculation of the Earth-Sun distance by Aristarchus—Eratosthenes's older contemporary, active during the 3rd century BCE—was based on correct geometrical assumptions, but because of its incorrect angle estimate it was off by a factor of 20. His measurement was only about 5 percent of the actual size, and Ptolemy put the distance between the Earth and the outermost fixed sphere of stars at just 20,000 Earth radii (or about 127.5 million kilometers).

During the 9th century CE, the leading Arabic astronomers al-Farghānī and al-Battānī used only marginally different values, and seven centuries later, in 1588, Tycho Brahe actually put fixed stars closer to the Earth in his model of the universe than did the previous estimates: just 14,000 Earth radii, shrinking the ancient cosmos by a third distance-wise and nearly three-fold volume-wise! Galileo's pioneering telescopic observations (their results were published in 1610) revealed the previously unseen multitude of more distant stars and made the idea of a highly constrained shell-like universe, imagined by medieval astronomers, untenable.[82] Just as Europe began to descend into the Thirty Years' War, Johannes Kepler published (between 1618 and 1621) his monumental *Epitome Astronomiae Copernicanae*, still beholden to the model of fixed stars but enormously expanding the size of the universe by setting the inner radius of the sphere equal to 4 million solar radii, or about 2.8 trillion kilometers.[83]

By the end of the 18th century a variety of astronomical observations made the true length of the astronomic unit (the distance between the Earth and the Sun—about 150 million kilometers) incontestable.[84] The next fundamental breakthrough in establishing the size of the universe came with measuring the distances between Earth and nearby stars, by tracking changes in their apparent position as seen from the Earth orbiting the Sun. Three pioneering measurements were published between 1837 and 1840, when Wilhelm von Struve, Friedrich Bessel, and Thomas Henderson measured the distances to three prominent stars—respectively, Vega (the brightest star in Lyra constellation), 61 Cygni (a double star in Cygnus), and α Centauri (the closest star to our solar system)—and modern analysis of their results has confirmed their accuracy.[85] Bessel's value was only about 10 percent smaller than the true distance that puts the star's distance from the Earth at about 11.4 light-years (one light-year = 9.461×10^{12} kilometers): even the nearby stars were trillions of kilometers away.

In 1917, Harlow Shapley, at Harvard College Observatory, offered the first size estimate of our galaxy, concluding that the Milky Way is about 100,000 light-years across and a few thousand light-years thick.

And not long afterward, in 1924, Edwin Hubble concluded that the Andromeda Nebula is not a part of the Milky Way, that it is a separate galaxy, and that this closest galactic neighbor is about 1 million light-years away (the actual distance is 2.36 million).[86] Subsequent progress of astronomic observations (now often carried out from satellite-based telescopes) pushed the size of the visible universe by four orders of magnitude. The farthest objects we can see are 13.8 billion light-years away, but because of the continuous (and initially rapid) expansion of the universe, the galaxies whose light traveled for nearly 14 billion years to reach the Earth are now about 47.8 billion light-years away, making the known universe about 93 billion light-years across.[87] Between 1600 and 2020, our observations and calculations expanded the radius of the known universe by 15 orders of magnitude—about 3.4 quadrillion times. Such truly astronomical numbers are not relatable to anything we encounter in our daily lives.

In contrast to some impressive astronomic calculations done during Greek antiquity, the journey in the opposite direction—into the ever-smaller worlds invisible to the unaided eye—began in earnest only during the latter half of the 17th century. Handheld magnifying glasses were in use for centuries, and hence it was the era's newfound curiosity and interest, driven by a greater propensity for experimentation and by more systematic scientific inquiries—as opposed to the relative ignorance of the Middle Ages—rather than any technical breakthroughs that opened the way.[88] The first magnifying compound devices (with one convex and one concave lens) were built during the early 1590s by Zacharias Janssen, a Dutch spectacle-maker. In April 1625, Giovanni Faber chose to call Galileo's compound design a microscope (Galileo called it an *occhiolino*, tiny eye), but it was only during the 1660s when microscopes were used for scientific observations.[89]

In 1661 came Marcello Malpighi's discovery of capillary vessels in frog lungs, and in 1665, after years of observations, drawings, and engravings, Robert Hooke published his *Micrographia*, the first collection of microscopic images.[90] The book was a great success, and it included much-reproduced plates depicting a flea, a fly, and flower-like stems of mold, as well as the first-ever illustrations of plant cells (of cork)—the name chosen by Hooke, as these partitions reminded him of

monastic *cellula*. The name stuck, and "cell" remains the scientific term
for the smallest unit capable of independent existence. Hooke's microscopes (built by Christopher Cock, a London instrument-maker)
magnified about 50-fold, but their images were poor, blurred, and much
affected by chromatic and spherical aberration. Consequently, the
impressive engravings in *Micrographia* were not reproductions of actually seen microscopic images of entire organisms, but reconstructions of
wholes assembled from many partial, and often imperfect, views.[91]

In 1677, *Philosophical Transactions of the Royal Society* published a letter
received from Antonie van Leeuwenhoek of Delft—a Dutch busi-
nessman, amateur scientist, and admirer of *Micrographia*—in which he
detailed his observations of life in droplets of water.[92] His simple

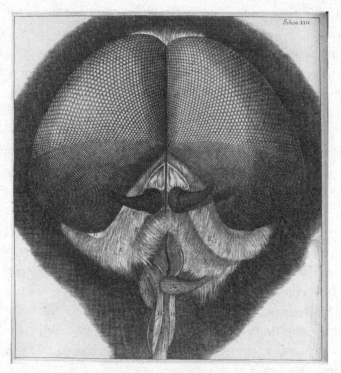

"I took a large grey Drone-Fly, that had a large head, but a small and
slender body in proportion to it, and cutting off its head, I fix'd it with
the forepart or face upwards upon my Object-Plate . . . I found this Fly to have,
first the biggest clusters of eyes in proportion to his head, of any small kind
of Fly that I have yet seen . . ." (Robert Hooke, *Micrographia*, p. 175).

single-lens microscope provided about 250-fold magnification, sufficient to investigate plant cells and powerful enough to discover a new world of "animalcules" whose existence was questioned by his contemporaries, especially as he concluded "that there were upwards of 1,000,000 living creatures in one drop of pepper-water," an infusion of both whole and crushed pepper in water that he prepared for his observations.[93] But this incredible claim was correct: van Leeuwenhoek was the first person to give detailed reports of both protozoa and bacteria. His microscope had a resolution of about 1.35 micrometers (1,350 nanometers, or about 1/30th of the diameter of a strand of human hair), good enough to see all but the smallest bacteria: large and ubiquitous *Escherichia coli* measures about 3,000 nanometers, but small *Mycoplasma* (a genus of bacteria that infects lungs, skin, and the urinary tract) has a diameter of only 200 nanometers.

Improvements in microscopy are best traced by looking at the maximum resolving power—that is, the ability of the lens to separate adjacent details of the observed objects. The best light microscopes cannot resolve parts that are closer than about 200 nanometers, or the size of small bacteria. Improved microscope lenses and designs reached that stage of precision early in the 20th century: we could see bacterial cells (with diameters between 1,000 and 3,000 nanometers) but only some of the largest viruses: *Pandoravirus* (1,000 nanometers, discovered in 2013) and *Mimivirus* (700 nanometers, identified in 2003) are exceptionally large. Most large viruses go up only to 250 nanometers; small ones are less than 30 nanometers in diameter.[94]

To see them required the next stage of microscopy, which began in 1939 when German physicist Ernst Ruska made commercially available the electron microscope (he received the Nobel Prize in Physics for this in 1986).[95] Early electronic microscopes had a resolution of no more than 100 nanometers, still not enough to see the smallest viruses. Gradual improvements enabled us to descend a few more orders of magnitude, and scanning electron microscopes now have a maximum resolution of about half a nanometer (the average diameter of an atom is 0.1–0.5 nanometers). Scanning tunneling microscopes (introduced in 1981; Gerd Binnig and Heinrich Rohrer were also rewarded for their efforts with the Nobel in Physics in 1986) have pushed the limit another order of magnitude, to just 0.01 nanometers.[96] Electron microscopes were first used to study metals, crystals, and ceramics; numerous challenges had to be overcome to examine living tissues.[97] We can now "see" a DNA helix (diameter of 4 nanometers) and even individual amino acids, constituents of proteins (0.8 nanometers).

This, then, is the size span we have determined using our ingenious technical means and theoretical calculations: from the diameter of a hydrogen atom (0.1 nanometers) to the diameter of the known universe (93 billion light-years). That's a dizzying difference of nearly 35 orders of magnitude, and it shows to what a truly unimaginable extent we have expanded our search for extremes in size.

II PERCEPTIONS, ILLUSIONS, MEASUREMENTS

To make sense of sizes, we must first perceive them.[1] Sight and touch are the usual senses we deploy, but sound and smell can also help in this effort. Sound as a measure of size can be a child's game: shouting in a large dark cave, then counting the seconds before you hear an echo, and multiplying by 170 because the speed of sound is about 340 meters per second. And in 2015 a carefully designed study showed that smell, a sense long associated with the homing pigeon, can also be used by people to orient themselves in a room: participants located a target in a room with higher accuracy when guided only by smell than in the absence of any smell, visual, or sound stimuli.[2]

But seeing—the dominant way of size gauging—is never a simple process. Our senses are embodied and, inevitably, how we perceive the world is determined by the attributes of our bodies: the position and orientation of our heads and trunks define and limit our perception of visual distance and object orientation. This takes place constantly and subconsciously. As the American geographer Yi-Fu Tuan put it: "The human being, by his mere presence, imposes a schema on space. Most of the time, he is not aware of it . . . In a literal sense, the human body is the measure of direction, location and distance . . . spatial prepositions are necessarily anthropocentric."[3] The quintessential proof of our very existence is our ability to move our bodies in expected ways and to maintain the sense of our being.[4] This truism is as obvious as it is profound. Gulliver's travels in Swift's novel are built on this body-based malleability (we will return to this book in the fifth chapter). What is normal for the adventurous fictional Englishman is seen to be monstrously large by the chipmunk-size Lilliputians. And on his next imaginary journey, the tables are reversed: what is normal in Brobdingnag is perceived by Gulliver as frighteningly large and grossly off-putting—the large spots, pimples, freckles of Brobdingnag ladies.

The size of our bodies and qualities of our senses determine how we perceive—constantly, subconsciously, or deliberately—the sizes of people, animals, objects, and landscapes. This also means how we appraise them (comparing specifics to internalized norms) and how we judge them (cherishing or discounting them), placing everything we see and value somewhere along that enormous spectrum of intuitively understood magnitudes, as well as along the size continuum that only modern science has made observable. When we look at a new face we do not think: *I will estimate the length of her central incisors and make a judgment about the attractiveness of her smile.* We do it subconsciously and automatically.

Humans are remarkably quick to judge. We decide about a person's attractiveness, likability, trustworthiness, competence, and aggressiveness almost instantly. Studies have looked into this: incredibly, a mere 100-millisecond (tenth of a second) exposure was enough to correlate highly with judgments that were made without any time constraint. Longer exposures produced more confident judgments, but they did not significantly affect the correlations between rapid and slow judgments.[5] For comparison, 0.1 second is faster than the time we need to trigger a simple physical action such as catching a falling object. And yet when you look at a face you judge its components immediately: you instantly compare the perceived crown length of a person's front teeth against the norm stored in your brain, and if they are even just one millimeter shorter than your "normal," you may find the smile less attractive.[6]

Similarly, you do not have to calculate the body mass index of a man (by estimating his weight in kilograms and dividing it by the square of his estimated height) waiting to board a plane in front of you. All you will need is a tiny fraction of a second to classify him as obese (a body mass index in excess of 30).[7] You do not need to weigh the meat and pasta portions served in American restaurants to conclude that they are much larger than similar meals in Italy.[8] And you do not need to measure the width and the length of a Japanese business hotel room to know that it is far smaller than the spaces offered by American hotel chains.[9]

And the physical limits of our bodies put obvious, and deeply internalized, limits on the sizes of many objects: tools must fit our hands, chairs must accommodate our posteriors. Many of their sizes can change only within very narrow confines, and this restricts new designs largely to aesthetic touches or to the use of new materials. Ubiquitous examples include glasses (about two-thirds of adults need to correct their vision), whose width is limited by the interpupillary distance; small home, office, and work tools (from screwdrivers to pencils), whose dimensions are limited by the size and proportions of human hands; sizes of cutlery, limited by the sizes of our mouths (or more accurately by the width of our buccal cavities); and dimensions of furniture (neither chair widths nor sofa heights benefit from enlargements beyond comfortable modular sizes).

Inevitably, all of this means that all sizes are perceived in relative terms as we compare them with our bodies, and as we use our experiences and expectations concerning the natural and designed environments we've encountered. Consequently, there are few experiences as memorable as those where we come upon unexpected sizes.

Expectations and surprises: preferred views and giant screens

You have been to Spain before, but never to the Basque country. And so you decide to visit San Sebastián, have a look at the great Bay of Biscay beaches and, en route to Santiago de Compostela, stop in Bilbao. You may not be interested in modern architectural design, but you read about Frank Gehry's sculpture-like museum and plan to check it out.[10] You leave your hotel, walk down Lersundi Kalea, then turn right—and across the street you see a building-size puppy in the shape of a sitting West Highland terrier (whose adult height is usually only about 25 centimeters) made of some 60,000 flowers planted in more than 50 tons of soil.

What are we to make of this oversize flower-puppy? Inevitably, some people love it, others are unsure what to make of it, some see it as too kitschy. But everybody is forced to react: the reversal of our expectations and the outsize reality demand it. This encounter is a fine illustration of our expectations and perceptions of size. Constant, cumulative experiences result in subconscious internalizations of expected dimensions, and encountering objects, images, and

Jeff Koons's floral Bilbao *Puppy* (1996, 12.4 meters tall).

experiences well outside such anticipated norms makes us react with surprise and, depending on the nature of these encounters, with emotions ranging from delighted approval to irrepressible fear.

Jeff Koons designed the massive (12.4 × 12.4 × 8.2 meters) sitting puppy in order, as he said, to instill confidence and security—to impart awe mixed with joy.[11] Even if you do not particularly like dogs or the idea of having an enormous steel-floral puppy, you must admit that the execution is admirable and that Koons—whose other objets d'art you might have seen and disliked, as I do his balloon dogs, his *Bouquet of Tulips* (a severed hand holding a bunch of aluminum flowers that look more like marshmallows than tulips), and his off-putting *Play-Doh* (an ugly pile of garishly colored aluminum chunks made to look like play putty)—succeeded in what he set out to do. Undoubtedly, the floral treatment is key—but so is the size: a tiny filigree puppy displayed in a glass case would be no more remarkable than the countless pieces of jewelry that can be admired in the display cases of every large museum.

Of course, encountering unexpected sizes is one of the key components of human perception. While many painters built their fame on

Trompe l'oeil paintings, such as Benvenuto Tisi's ceiling in Ferrara (1503–1505),
let us perceive three-dimensional sizes that are not really there.

an exact verisimilitude (from Francisco de Zurbarán's 1633 *Still Life with Lemons, Oranges and a Rose* to modern hyperrealistic paintings by Pedro Campos), others have resorted to deliberate exaggeration, distortion, or reduction. The first two choices are combined in Salvador Dalí's *The Persistence of Memory* (1931), with three malleable timepieces, and in his 1954 *Soft Watch at the Moment of First Explosion*, where a giant timepiece is, waterfall-like, oozing down a precipice while coming apart. René Magritte's 1938 *Time Transfixed*, with a steaming locomotive jutting out of a walled-up fireplace, is a fine example of a disorienting size reduction. And then there are completely illusionary sizes that are perfectly illustrated by *trompe l'oeil* ceilings—with two-dimensional paintings creating convincing illusions of three-dimensional space.[12]

But in everyday life we deal almost exclusively and repeatedly with well-known sizes—and, inevitably, expectations and context play a large role in our perceptions of these objects. After having seen them so many times, we have formed many standardized, normalized views of bodies, animals, machines, gadgets, and furniture. What kind of views do we prefer? When you try to imagine a teacup, do you see just its bottom (a rimmed circle, often with a maker's mark) or a straight-down view with dark- or amber-colored liquid confined by a circular rim? Almost certainly neither! In 1981, American psychologist Stephen Palmer and his coworkers tried to identify preferred viewpoints by resorting to three kinds of experiments: first, they showed people photographs of objects taken from different viewpoints (front, back, side, top, and 45-degree-angle views); then they asked them to describe their mental images of objects in terms of visible surfaces (again, front, back, sides, etc.); and then they used a camera to photograph objects from their preferred viewpoint.[13]

Their finding: regardless of the task, the participants consistently preferred the same off-axis viewpoints—above all, the three-quarters view that makes more than one surface visible. They called these preferences "canonical views," and concluded that these viewpoints

Wedgwood teacup seen from the top, bottom, and side—and its canonical view.

provide more information, as well as information of high importance. Further research confirmed that most objects are preferably seen from viewpoints 30–45 degrees above and rotated 30–45 degrees either to the left or the right. But is it that simple? At least four criteria may explain what makes a view canonical. How easily recognizable it is (it's hard to recognize a cup from a bottom view); familiarity (how often do we look at a cup straight down?); functionality (how do we see a cup when we hold it by its handle?); and aesthetic appeal (a downward three-quarter view is more pleasing than a straight perpendicular-sight view).

A canonical viewpoint as originally identified by Palmer and his colleagues is real and common enough—but it cannot be universal. We should not expect canonical views of hand tools to be the same as those of larger objects; there are differing preferences governed by overall proportions of the inspected objects. When three psychologists—David Perrett, Mark Harries, and Simon Looker—did experiments with real physical objects, rather than with images, the participants did choose off-axis views as the most informative. But (contrary to earlier results) when asked to rotate the objects to viewpoints best corresponding to their mental image, they preferred mostly "plan views"—that is, orthogonal lines of sight that do not convey the object's three-dimensional nature.[14]

These conclusions were confirmed by experiments done by Volker Blanz and his colleagues at the Max Planck Institute for Biological Cybernetics.[15] Off-axis views were clearly preferred in photographs: accidental views and views hiding important surfaces were avoided, and the choice seemed to reflect the maximal information content of such canonical views. Perhaps most convincingly, Elad Mezuman and Yair Weiss at the Hebrew University in Jerusalem turned to hundreds of views available on the Web's search engines and found that those images show nearly identical view bias (a preference for canonical views), as do laboratory experiments.[16]

At the same time, they found hundreds of counterexamples. These included canonical views of animals from animals' height rather than from a human viewpoint; off-center side views of cars, although we see them mostly from behind (as drivers) or from the side (as pedestrians); and views of tools as we use them—not, obviously, from a detached off-axis viewpoint. Often there is not a single canonical view but several preferred views, and when objects and creatures have unique features then those might be preferred: the unforgettable faces of some tropical monkeys, such as the Amazonian red-faced bald uakari, are excellent examples.

And in 2010, two researchers at the Massachusetts Institute of Technology, Talia Konkle and Aude Oliva, expanded the range of canonical designation from viewpoints to sizes.[17] Their experiments showed that objects are drawn, imagined, and preferentially viewed with a consistent visual size—and, remarkably, that this size is proportional to the decadic logarithm of the object's assumed size in the real world (for example, the logarithm for a 10-meter object is 1, for a 20-meter object it is 1.3), and that it is best characterized as the ratio of the object to the frame surrounding it. Not surprisingly, since it is analogous to canonical perspective, they chose to call this the "canonical visual size." Based on visual long-term memory, small objects were drawn small on the page, larger objects were drawn larger, and the drawn sizes were proportional to their size ranks and hence to the logarithm of their actual size.

In addition, drawings of smaller objects left more of the paper's space

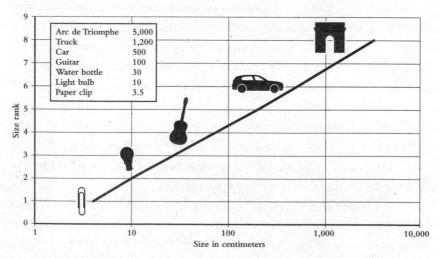

Arc de Triomphe	5,000
Truck	1,200
Car	500
Guitar	100
Water bottle	30
Light bulb	10
Paper clip	3.5

Two MIT researchers found a systematic logarithmic relationship between the actual size of an object and its size rank.

empty than did the drawings of larger objects, but small objects drawn on large paper were much larger than their real size. This leads to an intriguing conclusion that images of objects in our minds contain not only information about their relative visual size but also about a spatial envelope surrounding them. Drawings of objects are not scaled to the paper but to the object and its spatial envelope—which brings me to the sizes of frames and screens, and their impact on our perception.

In all premodern civilizations, the only way to see unexpected, larger-than-life human and animal bodies was to encounter monumental sculptures or oversize paintings. You can see the remnants of one of those statues (all that remains of Emperor Constantine's stone likeness: his head, arms, legs) in the courtyard of Palazzo dei Conservatori in Rome.[18] How do you react to a severed hand that is almost exactly 10 times longer than a normal male hand? Do you admire it and a similarly sized foot and head with bulging eyes as great sculptural achievements, or feel uneasy about their overpowering size? The intact statue would have been about 12 meters tall, just

Severed hand from the monumental statue of Emperor Constantine displayed in the courtyard of Rome's Palazzo dei Conservatori.

a bit more than a third of the height of one of the ancient world's seven wonders, the Colossus of Rhodes (33 meters), but that was a statue of a sun god (Helios), not a mere mortal, and it was brought down by an earthquake that struck the island just over 50 years after the statue's completion.[19]

And even the largest frescoes left behind by Renaissance masters, such as Vasari's *Battle of Marciano* in the Sala del Cinquecento of the Palazzo Vecchio in Florence, depict men and horses no more than 40 to 60 percent larger than their actual size.[20] Until the closing years of the 19th century, unless you visited such extraordinary sights, you could only see people and animals of normal size, or in reduced sizes in copied—and after 1450, printed—books and periodicals. All of that changed with the arrival of the Lumière cinematograph in 1895 in Paris and 1896 in New York.[21] Even the first American movie screen was 20 × 12 feet (6.1 × 3.7 meters), large enough to show human bodies twice their height and small animals or facial close-ups of truly colossal proportions. Of course, after lifetimes of exposure to screens of all sizes, we cannot now re-experience the disconcerting effects of such images.

What came after 1896 was rightly described by Erkki Huhtamo, a media historian, as a form of "Gulliverization" of the visual, as the unexpected has been reproduced in ever-larger and eventually also in ever-smaller screens.[22] The screen, an artifact that did not exist until 1895 (I mean the "passive" screen that, unlike the modern computer screen, would not allow any image manipulation by viewers), first became progressively larger and, eventually, its electronic form also became impressively smaller. The initial enlargement was required in order to display higher-resolution projected movie images, and soon after Lumière's invention it began its second and even more revolutionary advance—as an electronic display of images and texts. In 1897, Ferdinand Braun demonstrated the first cathode-ray tube, essentially an electronic gun firing images onto a phosphor-coated screen.

The Gulliverization of screens is now almost complete, close to the practical extremes in both directions. By 1905 movie screens were 20 × 25 feet (6.1 × 7.6 meters), by 1933 the first American drive-in was 40 × 50 feet (12.2 × 15.2 meters), and newly built IMAX theaters have screens up to 24.4 meters wide and 19.9 meters tall—even tall individuals can be seen 10 times their real height.[23] But outdoor projections on building walls dwarf that: the Empire State Building has displayed endangered animals 33 stories tall and fashion models 42 stories tall.[24] There is still some way to go: skyscraper-size images are not quite like Jules Verne's 1889 science fiction, where in the year 2889 newspapers are displaying mammoth advertisements on clouds.[25]

Moving in the opposite direction, television (starting slowly before the Second World War and spreading rapidly since the early 1950s) brought small screens indoors for the first time. Their sizes have followed the movie-screen trend of gradual enlargement, with today's preferred sizes having diagonals above 120 centimeters.[26] Since the 1970s, advances in microelectronics have enabled ever-smaller screens of portable computers and phones, and wrist-mounted screens of watches, monitors, and mobiles. Hundreds of millions of people are now watching videos and movies where human bodies are the size of tiny rodents or large insects. The smallest practical size of wrist-mounted gadgets is limited by the minimum number of pixels needed to get acceptable resolution: a 38-millimeter Apple watch (actual display size is 38.6×33.3 millimeters) has 340×272 pixels.

This, then, is the modern screen range: from about 4 centimeters to 150 meters, a 3,750-fold difference. And here is the Swiftian comparison: Lilliputians were 15 centimeters tall, Brobdingnagians about 21 meters, a 140-fold difference. Screens for watching images have gone far beyond Gulliverization; we can now see Lilliputians that are only about one-quarter of their Swiftian height—or project, on skyscraper walls, Brobdingnagians that are more than seven times taller than in Swift's imagination! For obvious size-based reasons—compatibility, convenience, portability—small screens have conquered the modern world. Just try to imagine today's everyday life without screens: the stationary ones permeate all modern industries, serving as monitors of tasks ranging from the generation of nuclear electricity to car manufacturing and preparing food (we do not find it strange that a baking oven has a screen), and by 2020 the global total of portable screens rose to nearly 15 billion.

As a result, many homebound individuals who live alone or keep minimal contact with other members of their family now spend far more time with miniaturized images of bodies (moreover, often with completely machine-generated bodies) than with actual human beings, and some of them may not see a person for days on end.[27] Habituation can make many things seem normal, but studies have little good to say about the effects of these new small-screen addictions on sleep, diversity of interests, and direct social interaction.[28] I will never forget the first time many years ago when I saw a family of four walking abreast on a city sidewalk, every one of them looking at a mobile screen. Today such behavior is unexceptional, even normal. This may be perhaps the most unexpected and the most consequential shift in our perception: how increasingly larger shares of us are spending more time in the world framed by small (computer) and tiny (mobile) screens, rather than seeing the world at its actual size.

Delusions of size and seeing what is not there

Our long history of size-driven expectations and surprises does not embrace only what is there, objectively and verifiably. Our perceptions are outcomes of habituation, expected regularity, and pattern expectation—but our perception of size is often subject to remarkable departures from reality. In order to see what is not there (or not to see what is actually there), there is no need to dream Alice-like or to have transient visions caused by some rare neurological disorder. There is no shortage of instances where we perceive sizes incorrectly.

The easiest way to illustrate this reality is by remarkable size illusions. Two impressive demonstrations were devised by Franz Carl Müller-Lyer in 1889, and by Hermann Ebbinghaus in 1902.[29] In the first, two equally long lines are terminated with arrows pointing inward or outward: lines with outward-pointing arrows are invariably seen as shorter than those with inward-pointing arrows. In the second, a circle surrounded by smaller circles appears larger than an equally sized circle surrounded by larger circles.

By far the easiest demonstration of a convincing size illusion requires only two sheets of standard A4 office paper.[30] Leave one sheet

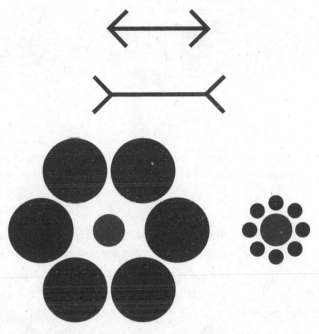

The Müller-Lyer illusion makes us misjudge the length of a line,
and the Ebbinghaus illusion does the same with the size of a circle.

intact, fold the other one in half—and then compare the two sheets by centering the halved sheet on the intact rectangle. Of course, you know that the larger sheet is twice as large (100 percent larger)—but experiments show that people perceive it to be only 64 percent larger, strongly underestimating the size of the white area. The illusion persists almost unchanged after aligning and centering the smaller sheet either to the shorter or the longer side of the intact rectangle, or placing it diagonally across it (viewers see the intact sheet only 67–70 percent larger)—and it disappears (all viewers see the larger sheet being twice as large) only after aligning the folded sheet to three sides at the bottom of the intact paper! Clearly, most people seem to be incapable of accurately comparing even two simple shapes that do not carry any distracting or potentially illusion-inducing markings. Or, as Claus-Christian Carbon, the experiment's author, rightly put it, there is a perfect "evidence of inability to perceptually integrate more than one geometrical dimension."

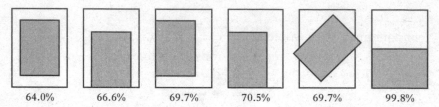

64.0%   66.6%   69.7%   70.5%   69.7%   99.8%

The white area is twice the size of the shaded area (hence 100 percent would
be a correct answer to how much bigger it is)—but participants in this simple
experiment consistently underestimated its size (the percentage values show the
mean estimations). Only the last presentation showed no significant difference.

And then there are subjective factors influencing size perception—
above all phobias, sensory deprivation (removal of stimulation), and
social status. There can be no doubt that fear influences the encoding
and processing of our observations. To arachnophobic individuals, tiny
spiders appear larger than wasps.[31] Acrophobic (fear of heights) individuals perceive a balcony to be higher than people who are not scared of
heights do, and they also overestimate object sizes when looking down
on them from an elevated spot.[32] In 1947, Bruner and Goodman showed
that children of lower socioeconomic status systematically overesti-
mated the size of coins when compared to the size estimates made by
children from much better-off families. This finding was recently confirmed by experiments using role manipulations (asking people to act as
if they were powerless or powerful). It appears that powerlessness leads
individuals to overestimate the size of valued objects.[33]

But phobias or deprivations are not prerequisites for widespread
perception errors, as most of us repeatedly fail even in simple tasks—
demonstrating that our quantitative grasp of size is surprisingly
limited. Because we constantly and subconsciously evaluate our surroundings, we do fairly well in relative classifications, qualitatively
assigning people, buildings, transportation, and machinery to basic
size categories of tall-average-small, long-medium-short, or big-
standard-mini. But people give notoriously erroneous answers when
asked to estimate areas and volumes. One of the most interesting
examples in the latter category is that people almost always underestimate the amount of liquid held in their mouth.[34] As a result, when

asked to ingest the same amount of liquid as is displayed in a transparent container, they would consistently ingest more.

We do generally better when appraising linear sizes—width of a table, distance to the nearest street corner, distances we have covered—but even these simple tasks lead to consistent and substantial errors. One of the most convincing estimation discrepancies that is readily demonstrable on any day a full moon rises in the evening sky is the so-called Moon size illusion. We perceive the Moon to be larger near the horizon than at the zenith of its arc. The effect was traced using a brain imaging technique to a critical node in the brain which deals with interpreting visual information.[35]

Extensive evaluations combined with field and laboratory experiments have demonstrated the ubiquity of distance misperceptions and identified some of the likely factors leading to such errors. To note just a few interesting findings, thanks to experiments in tumbling and tumbled rooms (realistically decorated spaces that can be changed to vary the relative orientation of sensory cues relative to upright position), we know that perceived distance depends on the orientation of both the body and the visual environment.[36] From experiments with estimating distances between points on computer displays, we know that the accuracy of these estimations declines when subjects have to estimate distances from memory.[37] The brain's representation of the arm modulates our depth perception and, remarkably, this internal calibration can be quickly adjusted in experiments with artificially enlarged body structure using computerized distortions.[38] And, not surprisingly, microgravity affects cognitive visual-spatial processing: crews at the International Space Station perceive objects' heights and depths as taller and shallower, and they also generally underestimate distances.[39]

Sometimes, these errors in estimating distances may actually be welcome. Vessels carrying people watching humpback whales off the shores of Maui should not approach the animals closer than 100 yards (91 meters), and a study showed that inexperienced observers erred by as much three-fold, thinking they were 100 yards away instead of 300 yards, and even the experienced men who thought they were 100 yards away were actually 120–180 yards.[40]

Other times, perception errors may have fatal consequences. Experiments did not show any consistent differences in estimating distances from shore to boat or vice versa, but in an open water environment, most people tended to underestimate shorter (less than 400 meters) distances and overestimate longer distances.[41] This has obvious implications when a capsized boat forces a person to judge an estimated distance against their perceived capacity to swim to safety.

But by far the most common cases of erroneous perception with potentially serious consequences are eyewitness testimonies. No conviction should be based solely on such "evidence": recollections of height or weight are no more reliable than other notoriously questionable witness statements, from body build to facial features.[42] Height and weight are, besides age, the most commonly noted attributes in criminal witness statements, but there are no generally accepted criteria to classify estimates as accurate. When considering a body height difference of 10 centimeters as acceptable, then an estimate of 170 centimeters would be equivalent to 180 centimeters—but the latter value would be well above the US adult male mean.[43]

A witness's own height and weight also matters: they are used as crucial references to judge the target person. Another notable tendency found by testing is general underestimation of the height of taller people and overestimation of the height of shorter persons. This tendency to estimate values close to the mean is certainly influenced by the mental image of the average height, which may lead the witnesses to stay away from extreme judgments. And, contrary to the commonly accepted 15-meter distance as a cut-off for acceptable identification, there is no such sudden threshold—as correct identifications decrease with increasing distance in general—and yet witnesses, despite being too far from a target to provide an accurate identification, still choose to guess.[44]

Things get even more interesting when we're fooling ourselves about ourselves, and about what we eat. Body-based comparisons—using one's own body as a scale to measure the external world—are unavoidable, but people's mental images of their own body, continuously revised through sensory inputs, are highly malleable. Experiments with virtual reality show how susceptible we are to these manipulations. People in large virtual bodies perceive objects to be smaller and nearer;

while people in small virtual bodies have the opposite perceptions.[45] Moreover, these effects can be replicated with hand sizes: when a virtual hand is large, objects tend to seem smaller, and vice versa; and, reversing the standards, the size of familiar objects can also influence the perception of hand size.[46]

People, especially older and overweight individuals, frequently overestimate their height and underestimate their weight. In an American study, older individuals and males over-reported their height more frequently than younger people and women, and African American women were more accurate in their self-estimations than any other group.[47] Obviously, such errors result in artificially lower estimates of obesity prevalence. Similarly, a Nigerian study found that just over 40 percent of obese participants perceived their weight to be normal. In this particular setting, the misperception might be aided by the fact that roughly as many respondents viewed large body size as desirable.[48]

Another recent study illustrated the two opposite unhealthy shifts in these body size misperceptions. When presented with a large selection of unique body images, the study participants consistently miscategorized overweight male bodies as being normal weight (even though they accurately categorized normal weight), and consistently miscategorized underweight female bodies as normal (even though they still accurately categorized normal female bodies).[49] This leaves only one conclusion: the perceptual boundary of what is normal has stretched at both extremes but with distinct gender bias—overweight male bodies and underweight female bodies are now perceived as "normal" within this expanded category.

But the most worrisome reality of these misperceptions is the fact that many parents underestimate the body size of their children by placing them in a lower—healthy—body mass index category than they actually belong to. Having parents who do not see the problem as even existing makes it more likely that such children will stay overweight or become obese, an undesirable outcome that may last a lifetime.[50] And many perfectly normal adolescents and young adults see themselves as too small or too heavy because we compare bodies and faces with internal templates that are increasingly based on images of faces and bodies encountered daily in mass media.[51] Not surprisingly, these perceptions shift the norm well below the population-wide mean, and they foster dissatisfaction, particularly among adolescents and young women.[52]

Anorexia nervosa and body dysmorphic disorder are the two most dangerous manifestations of this reality. These psychiatric disorders involve a distorted experience of an individual's physical appearance: they may share common underlying abnormalities in visual processing and perceptual organization, including exaggerated attention to detail and reduced processing of larger-scale features.[53] In the most extreme stages of anorexia nervosa, emaciated teenagers and young adults see themselves as "fat" (unacceptably overweight) and refuse to eat in order to regain weight. This condition affects annually about 8 out of every 100,000 people, with the highest average prevalence rates (0.3–1 percent) among young women.[54] This disabling, deadly, and costly mental disorder has been difficult to treat, though promising interventions include not only cognitive remediation therapy but also targeted brain stimulation.[55]

Individuals with body dysmorphic disorder are preoccupied with misperceived defects in appearance and see nonexistent or minimal flaws as unacceptable; in men this is often focused on increased muscle size, whose desirability has been propagated by widely circulated images of perfectly toned and muscled bodies.[56] This condition affects 0.7–2.4 percent of the general population. Unfortunately, many people seek to treat their misperceived bodies through plastic surgery, whose subjects run across the entire life span—from five-year-old children to octogenerians.[57] Imaging studies also indicate that both disorders are linked to abnormal activation in several of the brain's systems involved in visual processing.

Finally, a less depressing take on size perceptions, focusing on food: only simple stimuli suffice to affect our judgment of food sizes. A Dutch study showed that when dieting participants were primed by seeing a cover of a culinary magazine depicting a tasty dessert, they then estimated the size of a chocolate muffin to be larger than did the non-dieting participants.[58] Moreover, some three-quarters of American chefs think that they are offering "regular" portions, while the actual portions of the two popular choices, steak and pasta meals, were found to be two to four times larger than serving sizes recommended by government dietary guidelines![59]

Larger meal sizes served in restaurants contribute to larger bodies. Plates did get larger: some American restaurants have been serving 64-ounce steaks (1.81 kilograms of beef), or triple-triple burgers whose nine patties add up to 1.1 kilograms of meat. The area of a circle increases with the square of its radius, and hence the increase in size of a typical American dinner plate from a 25-centimeter diameter during the 1980s to a 30-centimeter diameter a generation later has resulted in a 44 percent increase in area.[60]

The logical expectation would be that people would load more food onto their larger plates, and using smaller dishware would seem to be an easy and effective way to help people to eat less and prevent weight gain. Some studies have confirmed the effect. Wansink and van Ittersum, two researchers of consumer behavior, showed that fill levels vary directly with plate size.[61] Chinese buffet diners using large plates served themselves 52 percent more food and ate 45 percent more of it (and wasted 135 percent more!) than those with smaller plates.

This would suggest that a preconceived visual plate-fill level (perhaps on the order of 70 percent) determines the size of self-servings, and the so-called Delboeuf illusion may offer the best explanation of this effect.[62] Or it may not: a meta-analysis of food intake during the experiments with smaller and larger dishware found that in most cases there was no significant difference—plate sizes had only a marginal effect, and this means that recommendations to use smaller plates may not be an effective way to tackle overeating.[63]

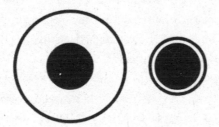

This old (created in 1865) illusion—an identical circle is perceived to be smaller within larger surroundings—seems to explain the desire for larger portions on larger plates.

Measurements: the advantages of being tall

Rich evidence indicates that perceptions—based on questionable, incomplete, or even illusory evidence—can be biased, misleading, and even perilous, and in order to deal properly with the countless consequences of size we must go beyond basic qualitative terms and quantify variables as accurately as needed. Antonyms we use are rarely neutral and are often loaded with feelings of satisfaction and unease, power and weakness, success and defeat, awe and relief: large vs. small, long vs. short, heavy vs. light, obese vs. skinny, deep vs. shallow, endless vs. limited, gigantic vs. minuscule, indefinite vs. circumscribed. In the absence of measurements, early societies had to operate only on a comparative (relative) basis, and one of the most telling divisions in human evolution is not just a gap but a veritable abyss between such societies and their successors—who were able to not only measure size routinely, but also use these measurements in order to derive universal rules and practices for geometry and algebra, as well as taxation, bookkeeping, and commercial exchange.

For millennia, physical measurements were done in the simplest possible ways: by pacing and by using body parts (thumbs, palms, feet, elbows), sticks, rulers, and ropes. Without these fundamental advances in measuring size, there would have been no wonders of the earliest civilizations: no clay cities and ziggurats of ancient Sumer (made from uniformly sized bricks), no Egyptian pyramids (made from stones chiseled to admirable tolerances), no perfectly proportioned Greek temples. Concurrently, these direct ways of basic measurement were supplemented by indirect approaches relying on basic geometry. These practices and procedures were simple, but the results are astonishingly accurate.[64]

The Great Pyramid of Giza is aligned north-south with an accuracy of 0.05 degrees, and it could not have been done by using Polaris (the North Star, α in Ursa Minor), which is now almost perfectly coincident (+89° 15' 50.8") with true north—4.5 millennia ago, no star occupied that spot, so the most likely explanation is that the builders watched two stars (Mizar and Kochab) whose vertical alignment produced the imaginary line passing through the North Pole.[65] The beauty of this solution is that it also gives us so far the best way to date the pyramid's construction: the precise alignment of the two stars took place in 2467 BCE.

And without accurate measurements, ancient Greek builders could not have incorporated the many departures from straight lines and right angles that combine to make the Parthenon aesthetically pleasing. The massive stone structure is not rigidly rectilinear.[66] The temple's stylobate is not flat but domed—its corners are lower than the building's center, in order to avoid optical sagging in the center, and the downward bent is replicated by the entablature, the superstructure of the temples. Columns of the surrounding Doric colonnade are leaning

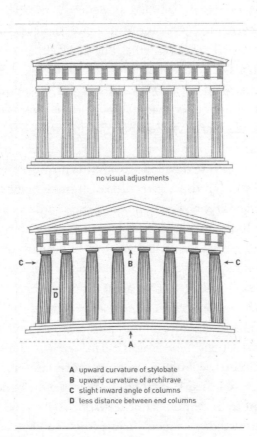

no visual adjustments

A upward curvature of stylobate
B upward curvature of architrave
C slight inward angle of columns
D less distance between end columns

Slight size adjustments, exaggerated in the bottom drawing, were incorporated into the Parthenon's design in order to achieve desired visual effects.

slightly inward; the corner columns have slightly larger diameters and lean inward in two directions, and the columns are wider in the middle. Despite these early achievements, measurements of size remained complicated by the absence of widely used standards. Romans imposed their measures on their far-flung empire, but even at its peak it occupied only a small fraction of the continental mass.[67]

Lack of standardization prevailed into the early modern (1500–1800) era, with not only countries but also cities using their own versions of linear, mass, and volume units. And it would not be an exaggeration to conclude that the post-1500 rise of western civilization to its eventual dominance of science and engineering and to its high economic efficiency and productivity was a direct result of ever more sophisticated, ever more widely applied (to energy use, to nutrition, to economic performance), ever more accurate and eventually also universally standardized measurements. The US is the world's only major country that persists in its use of outdated units.[68]

There is perhaps no better way to illustrate the importance of even simple size measurements than taking closer looks at height, that quintessential size that everybody can relate to. Systematic measurements of this fundamental size are surprisingly rare. Inquisitive Europeans were making detailed astronomic measurements for centuries before they began to keep records of human height.[69] And it was only in 1759 that a French savant pursued the simple idea of measuring the growth of human height from birth to maturity. Philibert Guéneau de Montbeillard was a lawyer, but like many of his educated contemporaries in pre-revolutionary France he was also a writer, naturalist, and a contributor to that great assemblage of enlightenment knowledge, the multivolume *Encyclopédie* edited by Diderot and D'Alembert. His son François was born in 1759, and until his 18th birthday in 1777 Guéneau de Montbeillard measured his height every six months. Comte de Buffon, an even more famous naturalist and encyclopedist, published the table of these height measurements in the supplement to his famous *Histoire Naturelle*.[70]

By his 18th birthday François had grown to 187 centimeters, far taller than his average contemporaries and as tall as an average teenager in today's Netherlands, the world's tallest male population of the early 21st century. It took a few more generations before the pioneering statisticians took their first close looks, during the late 1820s and the 1830s, at the height distributions of military recruits and the growth of children and adolescents (for details of that, see the sixth chapter of this book).[71] The first practical benefit of systematic human height studies was to provide the information about expected childhood development that began with the publication of charts of human growth of Massachusetts children in 1891.[72]

Growth curves—which are now consulted by family doctors and pediatricians to check if children thrive or lag behind, and hence to assure anxious mothers or to alert them to possible growth deficiencies—are based on large-scale measurements of height and weight in infancy, childhood, and adolescence. Some countries use their national standards; others rely on growth charts developed by the

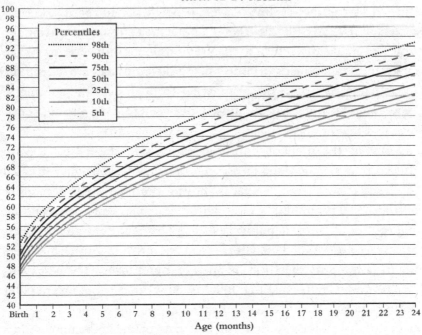

Actual size in centimeters
Birth to 24 Months

Length-for-age percentiles for girls from birth to two years of age, according to the World Health Organization.

World Health Organization.[73] This guidance is given within well-defined boundaries, and the numbers are highly reliable: at age two, a baby girl measuring 86 centimeters is perfectly average, but if she is 82 centimeters then only 10 percent of children in her cohort are shorter, and at 91 centimeters she would be exceptionally tall, in the 95th percentile.

The second benefit came from the study of measurements of human bodies in the past. This research began only during the 1960s, but by the beginning of the 21st century it resulted in some interesting reconstructions of the development of human height since antiquity, and in many more detailed and much more accurate histories of human height during industrialization in the 19th century, plus health and nutritional advances in the 20th century.[74] Even partial skeletal remains make it possible to determine body heights (femur length has the best correlation with body height), and we now have many historical reconstructions tracing long-term trends in European height.[75]

One perspective, with data extending for a millennium, was based on more than 6,000 skeletons from the northern part of Europe, and showed a significant decline of average heights from about 173 centimeters in the early Middle Ages to about 167 centimeters during the 18th century, with a return to the previous highs only after 1900.[76] In contrast, measurements of nearly 9,500 skeletons dated between the 1st and 18th centuries, and originating from all over the continent, show long-term stagnation, mostly between just 169 and 171 centimeters for adult males.[77]

Most of the modern European growth in stature took place between the 1870s and the 1980s, when average male height increased by about 11 centimeters—that is about a centimeter per decade. In northern and central Europe the gains were most rapid between 1910 and 1955, as advances in public health and hygiene negated the effects of two world wars and the Great Depression, while southern Europe had its fastest growth period between 1950 and 1980.[78] The best available reconstruction of Dutch heights shows that people were taller during the late Middle Ages and at the beginning of the early modern period than they were in the first half of the 19th century, just before a spectacular increase in Dutch heights began. This trend accelerated during the second half of the 20th century, making the Dutch the tallest nation in the world.[79]

Thanks to the mass-scale NCD Risk Factor Collaboration study, we now have an excellent understanding of the global realities of human height. The study re-analyzed nearly 1,500 populations—including almost 20 million people born between 1896 and 1996 in 200 countries.[80] During that century, women gained 8.3 centimeters and men 8.8 centimeters, with early linear increases followed by post-1950s plateaus. South Korean women had record gains, averaging 20.2 centimeters, and Iranian men got 16.5 centimeters taller. European and North American gains were smaller; Japanese gains were briefly reversed during the Second World War, but they still amounted to about 14 centimeters for the conscripts.[81] Chinese growth during the second half of the 20th century added just 6.8 centimeters for men and 5.4 centimeters for women, and sub-Saharan Africa and South Asia saw either minimal or no average height increases.[82]

Dutch men (average of 183.8 centimeters) now find their closest competitors in Montenegro (183 centimeters), Estonia (182.7 centimeters), and Denmark (181.9 centimeters), and on the smaller-scale level the world's tallest male population (184.1 centimeters) is now in Split-Dalmacija county in Dalmatia, the coastal part of Croatia. The Netherlands (170.4 centimeters) and Estonia (168.7 centimeters) also have the world's tallest women, while the lowest average male statures (less than 160 centimeters) are in Timor-Leste, Yemen, Laos, Madagascar, and Malawi. The shortest adult women (average of less than 150 centimeters) live in Guatemala, the Philippines, Bangladesh, Nepal, and Timor-Leste. And, remarkably, the century of unprecedented height gains did nothing to narrow the absolute difference between the world's tallest and shortest populations: in the year 2000 it was about as large (20 centimeters) as in 1900.[83]

The third benefit of systematic height measurements and analyses came from realizing and confirming a large number of desirable life indicators associated with height. As the anthropometric studies multiplied, it became clear that height—a simple and easily measured variable—is a convincing proxy for quantifying many aspects of human well-being, and a revealing marker of human welfare that correlates with many social and economic variables.[84] Tall people are perceived to be healthier, smarter, more confident, better educated, socially more adept, and better liked—and, on average, these qualities make them more accomplished, richer, and more influential.

Such claims demand closer appraisals and encounter inevitable complexities. Adult height is highly polygenic; many regions of the human genome have been linked with height, and studies of twins and adopted children suggest that up to 90 percent of all normal height variation within a population can be ascribed to genetic factors.[85] At the same time, genetic height potential can be realized only in the presence of a healthy living environment and in the absence of chronic ailments and nutritional deprivation during childhood and adolescence, and hence adult height is also the product of familial economic and social circumstances. Tall individuals are more likely to have been raised in healthy and wealthy settings; in turn, taller and healthier people can work harder, both physically and mentally, and realize higher incomes, and tall males have an easier job of finding partners.[86]

Advantages begin early, and parental education and household incomes made the greatest between-country differences in a study that compared the growth of children in five affluent countries. Taller children came from families with educated parents and households with higher incomes, but in more equal countries (Sweden, the Netherlands) children were taller at all levels of education and income than in the economies with less equal income distribution (the USA and the UK, for instance).[87] Robust growth in childhood and adolescence is linked with better cognitive function, better conduct of daily life, and better mental health.[88]

A convincing demonstration of this reality was provided by a study of disparities in children's vocabulary and height, in relation to household wealth and parental schooling, in four low-income countries—Ethiopia, India, Peru, and Vietnam. Children's linear growth and vocabulary were tested at ages 1, 5, 8, and 12, and those coming from the top quartile of socioeconomic status were taller and had better language command than those from the bottom quartile, with the difference in language scores being larger than in height scores.[89] Data on German preteen students showed that height's positive association with educational achievement is valid in affluent societies: even after controlling for academic achievement and family background, taller children were more likely to enroll in gymnasium (the secondary school track that leads to university).[90]

Moderately positive correlation of height and IQ was noted generations ago. Both of these traits are partly inherited, and studies of twins indicate that upbringing might be responsible for nearly two-thirds of the height-intelligence correlation, and genes for the rest.[91] Height's most consequential positive association is with income, because money in modern societies does buy better nutrition, habitation, healthcare, and education. New York University's professor Enoch Burton Gowin presented the first systematic review of the link in 1915 in his book on *The Executive and His Control of Men: A Study in Personal Efficiency*: bosses were taller than employees, bishops taller than preachers, sales managers taller than salespeople.[92]

Little has changed a century later. During the first two decades of the 21st century numerous studies have noted the height-wage correlation in affluent western nations, as well as in lower-income and rapidly modernizing countries in Asia and Africa—including China, Taiwan, India, Indonesia, and Ethiopia.[93] The link holds both for occupations requiring physical strength (where higher height and weight might be of obvious mechanical advantage) and for those consisting of intellectual tasks. Studies attributed this premium both to non-cognitive abilities correlated with stature and to cognitive advantages. Research using data from Britain's National Childhood Development Study demonstrated that taller children have higher average cognitive and non-cognitive test scores, and that each aptitude explains a roughly equal share of the later height-earnings premium.[94]

The height-income gradient is steep at low incomes (incomes rise faster for taller workers engaged in low-income jobs), but it flattens subsequently and levels off for annual incomes above $20,000 per year. And in low-income countries, a $1,000 increase in average income is associated with 0.8 centimeters of higher stature in men and a mere 0.3 centimeters in women.[95] In affluent countries, the positive link holds all the way to the top of the labor market, and a near-exhaustive examination of 28,000 Swedish men who acted as CEOs between 1951 and 1978 (as expected, a cohort taller than the population average) showed that their height increased with the company's size. Those managing enterprises worth more than SEK 10 billion averaged 183.5 centimeters, while those running companies worth less than SEK 100 million were more than 3 centimeters shorter (180.3).[96]

And height beats even the might of China's ruling Communist Party. A study of incomes in China found that, for males, every centimeter increase in height led to a 0.05 percent increase in the probability of becoming a CP member. Members earn about 11 percent more than non-member males and hence a centimeter increase in height brings about a 0.06 percent increase in hourly wage for party members (what the authors call "height premium in earnings through the political channel"). In contrast, when controlling for this height premium, a 1-centimeter increase in height brings 1.18 percent increase in male hourly wages, which means that the height premium via the market channel soundly beats the political advantage.[97]

But the most recent Chinese analysis that takes into account the individual genetic data of more than 3,000 individuals offers a conclusion that best captures the reality of the height-income link.[98] In this study, an additional centimeter in height is associated with a large (10–13 percent) increase in annual earnings, but height per se has only a negligible causal impact. Everything else being equal, an additional centimeter in genetically determined height raises annual salary by a statistically insignificant margin of less than 0.6 percent. Consequently, the observed height premium is not a reflection of labor-market discrimination based on height; instead, the observed height premium captures the impact of other, truly causal factors, including health, mental attributes, and risk preference.

Are there any downsides to being tall besides inadequate legroom (particularly in narrow-pitch airline seats: see the fourth chapter), tending to have bad posture, and, for exceptionally tall people, finding clothes that fit adequately and always having to tower above a crowd? Contrary to earlier studies linking increased height to longer life expectancy, the new consensus is that, given the same healthy diet and lifestyle, shorter and slimmer individuals live longer than the taller ones.[99] Comparisons of various ethnic groups in California showed a decrease in life expectancy per centimeter of height, ranging from 0.4–0.63 years; similarly, subsequent studies of life expectancies in Cuba, Italy, Poland, Spain, and the US found decreases of 0.5–0.7 years per additional centimeter of height.[100]

A recent study looked at nearly 4,000 living and deceased men who played in the National Basketball Association (debuting between 1946 and 2010) and in the American Basketball Association (starting between 1967 and 1976). Their average height was 197.8 centimeters (range 160–231.1 centimeters). The tallest players (in the top 5 percent) died younger than the shortest players (in the bottom 5 percent) in all but one birth decade, the 1940s, and survival analyses also showed an inverse relationship between height and lifespan.[101]

To a significant degree, these differences are due to the association between taller stature and higher risk of many cancers. Studies in Europe (5.5 million men and women in Sweden; a million women in the UK) and the most recent study in South Korea (involving a cohort of 23 million adults) found this relationship to be fairly linear, valid for both sexes. But for most cancer sites the magnitude of the association between taller stature and the disease is greater in women. And the link is valid for as many as two dozen different cancers, ranging from those of the nervous system, thyroid, and breast to lymphoma and leukemia—but oral cavity, urinary bladder, pancreas, liver, and stomach cancers have been less consistently associated with height.[102]

The best explanation of the height-cancer link is the direct-effect hypothesis: more cells provide more targets for cancer-causing mutations, and height is thus simply a proxy for the total cell number and for increased cancer risk. Data from four large-scale cancer surveillance projects (including 23 different types of the disease) confirmed this: there is about a 10 percent increase in cancer risk (12 percent for women, 9 percent for men) for every 10 centimeters of additional height.[103] As expected, this is not a perfect explanation: melanoma has an unexpectedly strong link to height, and unlike in women, height explains only about a third of the higher incidence of non-reproductive cancers in men, which means that other explanations must be considered.

Well-being and survival are much more imperiled by truly excessive growth. Untreated gigantism, a very rare condition caused by pituitary malfunction in children before the fusion of their epiphyseal growth plates (parts of long bones where growth is taking place), leads to premature death. Acromegaly—excessive pituitary activity after epiphyseal fusion—affects one out of about every 6,300 people, causes abnormal enlargement of bones (hands, feet, and face), and shortens lifespan but does not increase a person's overall height.[104] Marfan syndrome is only slightly more common (one out of every 5,000 people) and it produces tall bodies, long extremities, and frequent heart problems. A US president (Abraham Lincoln), perhaps the greatest violin virtuoso of all time (Niccolò Paganini), a renowned Russian composer (Sergei Rachmaninov), and a modern English poet (Edith Sitwell) were among those afflicted with this syndrome. As recently as the early 1970s, the medium life expectancy of Marfan patients was 48 years, with most premature deaths due to cardiovascular causes; by the century's end, life expectancy, thanks to cardiovascular surgeries and medical therapy, was significantly increased, to more than 70 years.[105]

We've looked at quantities ranging from sizes of objects to heights of children and adults, and at the ways we perceive—or mis-perceive—these realities. In the next chapter I will shift my attention to what I might label the qualities of size, to the proportions of objects and bodies (such as leg-to-body and hip-to-waist ratios), and to their symmetric and asymmetric structures and designs. These are particularly fascinating size-related topics. What rules in nature: symmetry or asymmetry? What body proportions were ideal in antiquity and to what extent have they changed in the modern world? Are symmetrical faces always seen as the ideal of beauty? Many scientific disciplines have contributed to understanding these questions better, and they have also helped to refute some persistent myths. Perhaps none of those has been more widely held than the belief in the extraordinarily common presence of one specific size proportion: the golden ratio.

III   PROPORTIONS, SYMMETRY, AND ASYMMETRY

All sizes can be measured or estimated more or less accurately in absolute terms. Major exceptions are the sizes of shapeshifting organisms, ranging from sluggish blobby slime molds to cephalopods, with some octopus species adding rapid changes in color to alterations in their elusive dimensions.[1] Remarkably, a recent discovery indicated that large shapeshifting organisms existed during the earliest era of complex life. These rangeomorphs—some only a few centimeters tall, others measuring up to 2 meters—lived during the Ediacaran period (635–541 million years ago), which was dominated by microbes and predated the Cambrian eruption of complex marine life.[2] Rangeomorphs had soft, fern-like bodies, and the indistinct evidence makes it impossible to be sure if they were a kind of fungi or if they were the first large-size animals. In any case, it appears that they did not make it into the Cambrian to share the ocean with trilobites and strangely shaped free-swimming creatures, including one actually named *Hallucigenia*, a tiny, bizarre-looking creature with long spines on its back.

All sizes can also be perceived in relative terms. We compare the overall size of organisms or artifacts to the sizes of their constituent parts, organs, or components, and where our bodies are concerned we are adept at noticing even slight departures from expected means— eyes too close or too far apart, head too round or too elongated, gait too short or strangely long. We judge sizes in relation to their natural or man-made settings and, again, instantly notice anything out of the ordinary: El Capitan's sheer vertical rise as seen from the bottom of the Yosemite Valley, or a cruise ship passing through Venice's Canale della Giudecca. The dizzying ascent of perhaps the world's most famous rock wall has been available since 2015 in the first-ever vertical Street View collection on Google Maps, and enormous cruise ships looming within the lagoon city were a common pre-pandemic sight.[3]

We follow the changes of sizes over time, wondering at the persistence of some forms and at the rapid evolution of others.

Proportions of organisms, landscapes, and artifacts are the key determinants of visual appeal as they guide our aesthetic preferences for human bodies (and particularly for faces). Necessarily, artists have been their keen observers. Albrecht Dürer, the preeminent painter of the German Renaissance, is just one of many outstanding examples: he produced copious drawings of human proportions that were collected posthumously by his wife and a friend in *Four Books on Human Proportion*—here are but three of them.[4] They have also studied natural phenomena (from landscapes to seascapes) at many scales, and the understanding of functional proportions is essential in designing an ever-increasing variety of machines and objects.

Proportions of hominin bodies have changed as evolution distanced them from those of our primate ancestors, and tiny sculptures of

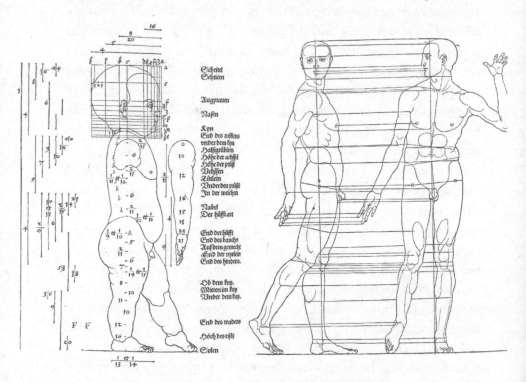

Bodies from Dürer's posthumously published
Vier Bücher von menschlicher Proportion (1528).

human bodies (some—so-called fertility figures—with exaggerated features) are among the first preserved artistic expressions, going back more than 20,000 years. This fascination has endured, and the proportions of our bodies have been studied extensively by life scientists as well as by visual artists, and some of their interesting observations and conclusions will fill the next section.

Proportions: bodies, buildings, paintings

Sometimes, against all advice about avoiding any boring openings, it is imperative to start with clear definitions. "Proportion" is the way to describe how one part relates to another, or how a specific part relates to the whole.[5] "Harmonious" is the adjective commonly used as a qualifier, but proportions can be anything but harmonious. There is also a restrictive algebraic definition of proportion: when the two ratios (or fractions) are equal (for example, $1/5 = 5/25$). Both definitions will be useful here.

The proportions of complex organisms and of many artifacts are dictated by their function, as well as by many genetic and environmental constraints. The proportions of our bodies are the outcome of several million years of evolution following the separation of hominins from great apes. Changes of body proportions during that time have been documented (with different degrees of confidence) from archeological records, as well as from recent studies focusing on time spans as short as three to five decades. The evolution of body size and body proportions among the early hominins was complex.[6] Sizes of *Australopithecus*, a genus of our early African predecessors, varied enormously in both space and time but, in comparison to our species, they were notable for their long bodies and short legs. *Homo ergaster* (early African *Homo erectus* who lived around 1.8–1.5 million years ago) was the first species with more modern body proportions— notably longer legs, allowing a faster walk.

At the same time, archeological records show a significant increase in the distances over which these hominins carried their stone tools. This means that, besides walking endurance, effective load-carrying might have been a factor in the evolution of modern body proportions.[7] Early

Homo fossils are dominated by relatively smaller individuals, implying that the emergence of our genus was not linked either to an increase in body size or to unprecedented increases in size variation.[8] So far, the most detailed analysis of body mass and statures of more than 300 hominin specimens (dating from 4.4 million years ago to the Holocene) demonstrated complex evolution patterns of body size variations: the phases of relative stasis are interspersed with periods of rapid increases. Early *Homo* retained considerable size diversity, while in later *Homo* the stature has been maintained from about 1.6 million years ago; but consistently higher body mass is evident only since about half a million years ago.[9]

Not surprisingly, climate has been a key influence, and a recent analysis has confirmed that human stature and body proportions conform to rules formulated by the German biologist Carl Bergmann in 1847 and Joseph Asaph Allen, an American zoologist, in 1877.[10] Bergmann's rule posits that the body mass of mammals increases in line with distance from the equator, and hence with decreasing temperature. Allen's rule predicts that warm-blooded animals in cold climates should have lower surface-area-to-volume ratios than those in warmer environments, and that this adaptation will produce shorter limbs and appendages.

This is supported by comparisons of mammals whose range extends from the tropics to the Arctic, and the confirmation of Allen's rule in modern human populations (species adapted to cold climates have shorter body appendages) also helps to explain some past adaptations in extinct hominins. Because of their shorter limbs, Neanderthals required more energy for walking, but measurements of metabolic needs indicate that shorter legs would have been advantageous because of the reduction in heat loss.[11] The most obvious examples in modern human populations are, on one side, the Arctic Inuit—with shorter limbs, relatively broader trunks, heavier bodies, and less prominent noses—and on the other, the comparatively taller, longer-legged populations in parts of Africa.[12] As expected, there are departures and exceptions, but the rule holds well enough after taking into account the hypothesized pattern of the diffusion of modern humans out of Africa.[13]

People in hotter and more arid climates have longer legs, providing a larger area for evaporative heat loss. While in humid conditions (where evaporative heat loss is ineffective), smaller body mass and hence reduced metabolic heat output is the primary way to cope with heat stress. This is convincingly illustrated by comparing different foraging populations that have remained largely unaffected by modern nutrition and health-care. Heights of !Kung foragers living in the desert environment of Botswana average 160 centimeters for men and 150 centimeters for women; while among the Efé people of Congo's Ituri rainforest, the mean for men is 143 centimeters, and for women 136 centimeters.[14]

Of course, optimizing thermoregulation is only one of several fac-tors determining human body size and its proportions. Genetic studies have identified hundreds of genes controlling growth and adult size variation, and there are also obvious differences caused by hormones, nutrition, and disease. Some populations inherited reduced sensitivity to human growth hormone, and thyroid hormone activity (induced by differences in day-length: short day-length increases it) has an expected impact on average height. This hormonal effect was convincingly dem-onstrated by prefecture-level studies in Japan. Short day-length increases the activity of the thyroid hormone and contributes to increasing height, and vice versa. As a result, residents of Hokkaido, Japan's northernmost island (latitude >40°N) are taller than people in its southernmost territory, Okinawa (26°N).[15]

In some hunter-gatherer populations, the selection for smaller size might have been driven by limited resources, while the extraordinar-ily high milk intake of East African pastoralists contributes to their height. In contrast, the combination of disease and poor nutrition in childhood leads to stunting—an undesirable condition still prevalent in many poorer parts of the world. And even under the best circum-stances, the combination of genetic and environmental factors may not produce what we consider to be a harmonious appearance in a variety of species. Among the most notable examples of this are the bodies of sperm whales, with their massive heads and disproportion-ately weak lower jaws.

As with so many other intellectual and mathematical firsts, rules con-cerning canonical (generally acceptable) proportions were set down in ancient Egypt and then perfected in ancient Greece. In the western world these notions of ideal adult body proportions, immortalized in sculpture, lasted until the arrival of abstract paintings and sculpture in the early 20th century—the art forms that did away with any set guidelines regarding proportions. Instructions provided by Polykleitos, a Greek sculptor who worked during the second half of the 5th century BCE, have been lost, but he used them to make his—later much-copied—*Doryphoros* (Spear-Bearer).[16] With head height (from the skull's top to the chin) as the unit, the adult male body should be eight heads tall, and sole-to-groin distance should equal groin-to-head distance.

Perhaps no other image of human proportions has been reproduced as frequently as Leonardo da Vinci's *Le proporzioni del corpo umano secondo Vitruvio*, a naked man fitting into a circle formed by legs apart and arms slightly raised.[17] Marcus Vitruvius Pollio, the Roman architect and engineer known for his writings, left a very detailed description of body proportion in his *De architectura*, and here is just the opening part:

> For the human body is so designed by nature that the face, from the chin to the top of the forehead and the lowest roots of the hair, is a tenth part of the whole height; the open hand from the wrist to the tip of the middle finger is just the same; the head from the chin to the crown is an eighth, and with the neck and shoulder from the top of the breast to the lowest roots of the hair is a sixth; from the middle of the breast to the summit of the crown is a fourth.[18]

Leonardo corrected the Vitruvian claim about the navel as the body's center and placed it, as did Polykleitos, in the groin.

A large number of ratios can be calculated using body size and the sizes of major body parts, including height-to-torso, height-to-arms, height-to-legs, or by comparing body parts (chest-to-waist, waist-to-hips, hand-to-fingers). I will take a closer look at just three of these possibilities: the leg-to-body ratio (LBR), with leg length measured from the perineum to the sole; the waist-to-hip ratio (WHP); and the ratio of the second digit of the hand (index finger) to the fourth digit (ring finger), commonly abbreviated as 2D:4D. The first two proportions have been widely used in painting and sculpture striving for harmonious proportions, and by researchers

Leonardo da Vinci's pen-and-ink drawing of a Vitruvian man, made around 1490.

trying to elicit preferences for the ideal male and female body proportions. The third choice may seem to be strange—but recently many more studies have been published on the implications of 2D:4D ratios than on any other human body proportions, and we'll see why.

In the western world, height has always been one of the markers of attractiveness, and relatively longer legs have always been preferred to longer torsos. The widely preferable leg-to-body ratio in famous paintings of western women has been close to 0.5 (legs accounting for half of the body length): Botticelli's two most famous female figures— Venus (*Nascita di Venere*) and Flora (*La Primavera*)—as well as Dürer's slender women have exactly that LBR. But a higher ratio has been one of the markers of post-1950 fashion-magazine beauty.[19] The legs of the tallest fashion models account for as much as 61–67 percent of their (obviously always above-average) height. In studies presenting images or silhouettes of women and men with normal, elongated, and shortened legs, both genders perceive shorter-than-average legs to be less attractive. A higher-than-average LBR is widely preferable, but excessive leg length decreases the overall appeal, and both very short and very long legs might signal undesirable genetic or health conditions, while short legs may also be an indicator of stunting in childhood.[20]

These conclusions were confirmed by a study including perceptions from 27 nations in Africa, Asia, and Latin America, as well as Canada.[21] Participants were presented with silhouettes, the first one with an LBR of 0.515 and then others with elongated or shortened legs. For women, those with legs shorter than the original were not preferred anywhere, and the most favored proportions of a female body were either the standard length or legs 5 percent longer (LBR 0.541); the latter ratio was the highest mean choice in 17 of the 27 nations, and the nations with the highest preference for that LBR included Spain, Malaysia, Jordan, Romania, Croatia, and Indonesia. Similarly, normal and slightly above-normal LBRs were preferred for men.

American women (not included in this study) were also found to prefer a male LBR at about or slightly above the mean (and arm-to-body ratio being of no consequence), while in Japan values close to the mean were preferred by both sexes.[22] In contrast, a study of British undergraduates revealed a preference for a very high female LBR (about 0.7). The same authors found a very different outcome in Malaysia, where medium female and low male LBRs were the most attractive choices.[23] The finding for Malaysian females is in contrast with the outcome of the 27 nations study, a good reminder about the questionable reliability of investigations with relatively small numbers of participants.

There are also interesting findings about leg length and human health. Disproportionately short legs have been linked to enhanced health risks, from having a higher probability of coronary heart disease—primarily due to high blood cholesterol levels, impaired glucose and insulin regulation, increased blood pressure, and higher levels of blood-clotting.[24] Undernutrition in childhood not only reduces adult leg length relative to total stature, but it may lead to impaired metabolism: that would explain correlations between leg length and higher risks for excessive weight and obesity, diabetes, hypertension, low bone density, and other health problems.

But childhood undernutrition does not explain why children with shorter legs are at greater risk of disease and mortality at earlier ages. In women, short legs have one unexpected functional correlation: age at first menstruation is associated with stature, primarily with leg length, but the effect varies across populations and time. Among more than 3,000 American women (21–40 years old), earlier menarche was associated with shorter leg length (not with overall stature or sitting height), and almost two-thirds of the effect was due to the lower (distal) leg.[25]

In contrast to the leg-to-body ratio, whose desirable values have become more widespread thanks to better nutrition and the almost complete elimination of stunting in all affluent countries, the waist-to-hip ratio (WHR) has seen major deterioration. And this conclusion is not based on comparisons with the artificially reduced European waists of the 17th to 19th centuries. Those were created by tight lacing and corsets and were rarely as narrow as they appeared, because clothing designs helped to exaggerate the hip-waist contrast by using first farthingales and later wide crinolines or bustles.[26]

While adult bodies with waists of lower circumference than the hips remain the most desirable goal of human development, this outcome has become increasingly less common as post-1970 waves of obesity extended beyond the western world and became of major concern in countries ranging from China to Saudi Arabia.[27] In countries where more than half of the population is overweight and obese, the typical waist-to-hip ratio has seen substantial upward shifts, even inversions. According to the World Health Organization, normal WHRs are 0.9 and less for men and 0.85 and less for women, with high health risks associated with even a fairly slight increase among women (WHR>0.86) and a relatively larger increase in men (WHR>1.0).[28] A large-scale American analysis of people in their seventies showed no association between WHR and body mass index or waist circumference, but a significant increase of health risk with higher WHR.[29]

Changing WHRs are just one indication of the shifts toward heavier bodies in almost every age category. Because of its very large number of observations (nearly 14 million), a study of 19-year-old Germans conscripted between 1956 and 2010 is one of the best examples of relatively rapid recent body changes. Its results make clear that during that period the bodies of German conscripts changed from growing in height to growing in breadth. Height increased by 6.5 centimeters (to 180 centimeters), but this increase has plateaued since the 1990s; while the increase in average weight has continued, resulting in a higher prevalence of overweight and obesity.[30] Interestingly, the country also has both a north-south and a west-east gradient in height: in 2002, men and women from Schleswig-Holstein (just south of Denmark) were, respectively, 2.5 and 1.6 centimeters taller than their counterparts in Bavaria; and West Germans were about 1.5 centimeters taller than East Germans.[31]

And now, from macro features (LBR and WHR), to small-size proportions: looking at the index-to-ring-finger ratio. Modern interest in this ratio began in 1998 with the publication of a paper by John Manning and his colleagues at the University of Liverpool.[32] They anticipated that the pattern of digit formation is related to hormonal concentrations, and they showed that the so-called 2D:4D ratio in humans was different for men and women. Men tend to have a longer fourth digit (2D:4D = 0.98), while among women the two digits are of equal length (ratio of 1.0).

Interest in these findings developed slowly, but eventually the idea of a single number providing many possible insights into human health, disease, and behavior proved increasingly appealing. PubMed—the largest online archive of medical publications, containing some 32 million citations—makes it possible to trace the rise of this ratio in medical literature. 2D:4D papers took off in the early 2000s, rising from 51 papers published in the year 2000 to 150 papers a decade later, and to 283 papers in 2020.[33]

Recently published papers have looked into the links between the 2D:4D ratio and many diseases (from high blood pressure, ischemic stroke, and adrenal hyperplasia to gastric cancer, schizophrenia, and depression), behavioral problems (aggression in children, postoperative pain perception), and general abilities (with links ranging from hunting success among African foragers to the performance of female Olympic athletes).[34] Some researchers consider the ratio a reliable marker of fetal exposure to androgens (hormones regulating the development of male characteristics) and use the 2D:4D ratio to find medical and behavioral correlations.

But all of these conclusions rest on weak foundations. The difference between male and female ratios is small: in a large British sample, the average right-hand values were 0.984 in men and 0.994 in women, and the means vary among different populations.[35] As the studies accumulate, so does the doubt. Many biologists, physiologists, and endocrinologists have concluded that there is no strong evidence for testosterone involvement, that using the ratio is statistically problematic, and that many studies cannot be replicated (an increasingly common problem in modern science!).

As always, correlation does not confirm causation, and in these cases it also does not explain more than a small-to-moderate amount of variation for any examined links. Moreover, the ratio's sexual dimorphism may be largely the function of men's larger hands.[36] And while studies with female dog puppies showed that enhanced activity of androgen receptors produced the expected lower 2D:4D ratios, experiments with mice produced opposite results.[37] As a result, the journal *Hormones and Behavior* is no longer accepting any 2D:4D papers—but, as is often the case across science, given past interest in this topic the publication of studies in this questionable field will surely continue elsewhere.[38]

Cosmetic surgery is used to change the proportions of various body parts, and getting perfect teeth has become much easier than undergoing augmentations, reductions, or reconstructions of breasts, tummies, buttocks, noses, and chins, to say nothing about the complicated procedures required to lengthen legs. Proportions and ratios in organisms are products of evolution, often affected by malnutrition and disease, sometimes by accidents—but in art they are the key constituents of creations deliberately designed to please or console, to shock and offend.

Looking at how proportions have been deployed in oil paintings, classical and modern, is quite instructive, because this art form became, a long time ago, much more common than sculpture. Indeed, owning original paintings was one of the signs of new affluence during the early modern era: during the 1660s, urban households in Holland (whose population was at that time fewer than 2 million people) owned about 3 million paintings.³⁹ And until the beginning of the 20th century, "correct" proportions were expected by buyers and critics alike, and any deviation from this kind of realism was generally met with disdain.

Only a small minority of Old Masters departed from these dictates, by painting elongated forms: El Greco's Christ, saints and mythical personages are the best examples. In Parmigianino's great Uffizi canvas, both Madonna's neck and her son's body length are visibly unnatural. And during the 20th century, no other painter built his fame on such a limited color palette and on the elongated heads, necks, and bodies of his female models as did Amedeo Modigliani.⁴⁰ Still lifes had to be depicted in proper perspectives, even better if they deployed the tricks of *trompe l'oeil* and so achieved impressive three-dimensional effects on two-dimensional canvases.⁴¹ Tilting the perspective and flattening the composition, as Cézanne did in 1887 with *Still Life with a Peach and Two Green Pears* and as Georges Braque and Juan Gris were doing routinely after 1900 with their numerous planar still lifes, was unthinkable.⁴² Landscapes had to be finished in detail, not merely suggested by Impressionists—who, as contemporary critics complained, could not draw.⁴³

Despite the variety of styles, pre-19th-century paintings share a

El Greco's *St. Jerome as Scholar* (1609) and Modigliani's *Madame Zborowska* (1918).

"finished" aesthetic and adherence to proper proportions, but do they share some preferred or unusually common sizes and ratios of their canvases? The answer is clearly no, those variables have always been quite diverse. But because of our binocular vision field, the dominant direction of eye movements is horizontal, and hence it is not surprising that we find it easier to process horizontal images. Studies have shown that we identify people best based on the horizontally oriented information contained in the face image: Valérie Goffaux, a Belgian psychologist, concluded that "the human face is a horizontally-structured stimulus, and the processing of identity is tuned to the horizontal range."[44] Human preference for horizontal image processing has been confirmed experimentally by setting up horizontal and vertical store displays filled with identically sized bins containing the same number of items. Retailers must consider this reality when displaying their wares. Horizontal displays are easier to process (because of the match between the horizontal human binocular vision field and the dominant direction of eye movements required for their visual processing)—and because people see more variety in such displays, they process a horizontal assortment of goods more extensively.[45]

Knowing about the dominance of horizontality, it is not surprising that this orientation is particularly prominent in landscape paintings of all periods. This has been shown most convincingly by analyzing a large number of paintings—nearly 15,000 of them—representing the western tradition.[46] Before 1850, the vast majority were characterized by a horizontal-vertical (H-V) type of composition (example: a road or a meadow in the lower half of a painting, a clump of trees on the right), followed in frequency by horizontal-horizontal (H-H) designs (example: a horizontal line of bushes in the foreground, then a distant horizon separating land and sky)—but after 1850 the H-V type continuously declined while the double-horizontal (H-H) partitioning of rectangular canvases became the most frequent choice. Only abstract paintings thwart any attempts at discerning orientation and proportions. When 18 observers were shown 40 abstract canvases in the four listed combinations of horizontal and vertical orientations on a screen, and were asked to select the most attractive or the most meaningful orientation, only 48 percent of their choices matched the painters' composition intent.[47]

As for the proportions and the sizes of the world's most famous canvases, the Louvre's—and perhaps the world's—best-known painting, Leonardo da Vinci's *Mona Lisa*, is a small upright (portrait orientation) rectangle (77 × 53 centimeters).[48] Many other paintings that are routinely classed as the most famous or the best known, including Picasso's *Guernica* and Botticelli's *Birth of Venus*, are much larger, landscape-oriented rectangles.[49] But Klimt's *Kiss* in Vienna is a square with sides of 1.8 meters. Picasso's epoch-marking *Les Demoiselles d'Avignon* is nearly square (2.44 × 2.34 meters; 1.04 ratio), and so is the Prado's top attraction, Diego Velázquez's *Las Meninas* (3.12 × 2.76 meters, ratio 1.13), while Rembrandt's *Night Watch* (4.37 × 3.63 meters) in the Rijksmuseum in Amsterdam is only slightly more rectangular, with a ratio of 1.2.

Could we ever determine to what extent any specific shape of the framed canvas or its overall dimensions help to determine the outsize appeal of these paintings—or is it their subjects, their execution, and their uniqueness that are solely responsible for the prominence of these famous creations?

Symmetries: everywhere we look?

The Ancient Greek word συμμετρία (*symmetría*) did not have the modern, essentially mathematical, meaning of two matching halves. The original meaning was broader, describing harmony—well-proportioned, well-balanced arrangements bound up with beauty. Vitruvius, the most famous Roman engineer and architect, captured this essence: "Proportion is a correspondence among the measures of the members of an entire work, and of the whole to a certain part selected as standard. From this result the principles of symmetry. Without symmetry and proportion there can be no principles in the design."[50]

But even when understood in its more restrictive, mathematical meaning, symmetry is a ubiquitous universal property. There are many astrophysical symmetries, and our galaxy, the Milky Way, appears to have two arms extending from the opposite ends of its central cluster of stars and winding around its outer region.[51] Symmetries in nature range from orderly arrangements of atoms in molecules and the hexagonal symmetry of snowflakes to the bilateral bodies of mammals.[52] Many invertebrates display impressively complex symmetries: the skeletal structures of marine radiolarians (tiny single-celled organisms) conform to regular three-dimensional shapes with different numbers of flat faces, from octahedrons (8 faces) to icosahedrons (20 faces).[53] Similar ranges, from simple to highly complex, apply to symmetrical human designs, from the geometry of building foundations to admirable ornamental tile and ceiling patterns, carvings, and lattices in Muslim and Chinese interior design and architecture—whose three-dimensional counterparts are found in the arrangement of atoms in crystals.[54]

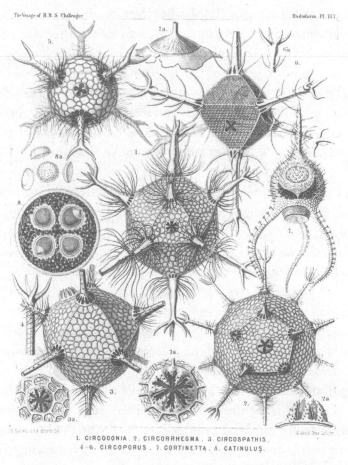

Radiolaria collected and drawn by Ernst Haeckel during
the voyage of HMS *Challenger* (1873–1876).

Symmetry's frequent manifestations in the structure of the universe, in the organization of terrestrial organisms and human bodies, and in our artifacts have been reflected in publications that not only explain its ubiquity and illustrate its complexities, but call it the building plan of nature, the universal principle permeating the entire universe, or the superprinciple of science and art.[55] But as one of the leaders of modern symmetry studies reminds us, symmetry's applications should not be overextended: "symmetry encompasses much—but not quite all."[56]

But, undoubtedly, symmetries are all around us. The most common regular geometric shapes are symmetrical—only the number of their lines of symmetry differs: a single line for an isosceles triangle; two axes for a rectangle, ellipse, rhombus; three lines for an equilateral triangle; four axes for a square; as many lines for regular polygons as the number of their angles (five for a pentagon, six for a hexagon, etc.); and an infinite number of lines of symmetry for every circle. Radial symmetry—with three or more divisions into sections whose rotations produce matching sizes and shapes—is dominant in flowers and fruits, although orchids, among the most spectacular of all flowers (and also less showy snapdragons and sweet peas) have simple bilateral symmetry. Bilateral symmetry, with shapes being mirror images along a midline, is the most common external pattern among animals (internal organs, however, are placed asymmetrically).

Among invertebrates, only sponges are not built on any symmetrical plan; all cnidarians (polyps, jellyfish, sea anemones) have radial symmetry; molluscs, arthropods (insects, crustaceans, centipedes, etc.), and all vertebrates have bilateral symmetry, with only some subtle, or hidden asymmetries. Differences in facial features, which hand we prefer to use (very few people, no more than about 1 percent, are ambidextrous), and the placement of internal organs are the most obvious examples of asymmetry in humans.

Leg dominance is much less appreciated than the obvious handedness, but it has some important consequences. A systematic review found clinically important differences, with the dominant leg performing better during the tests of quadriceps and hamstring strength and of single-leg hops. Leg dominance plays a role in the origin of some ligament injuries: female recreational skiers and soccer players are more likely to injure their non-dominant leg, while for the same sports the reverse is true for males.[57]

By far the most studied, and most disputed, occurrence of symmetry in humans concerns our faces. Of course, symmetry alone does not determine perceived beauty; facial proportions are no less important and their ideals are not immutable. Patients who were to undergo orthofacial surgery were asked to compare photographs of contemporary top-ranked beautiful male and female faces (drawn from annual *People* magazine and *FHM* magazine polls) with pictures of classical sculptures of 7 Greek goddesses and 12 gods. Their choice: an ideal female face is shorter than the male's—the opposite of antique ideals; a harmonious contemporary male face is longer than in antiquity; the ideal height of the lower face is 45 percent of the total facial height in idealized females and 48 percent in idealized males, compared to 50 percent for both genders in antiquity. And the ideal ratio of chin height to total lower facial height is 70 percent for both genders, compared to 66 percent according to the classical canon.[58]

The obvious problem of such comparisons is that the classical forms were based on ancient standards preferred by a specific and small population, and hence it is not surprising that their attractiveness has not been strongly confirmed by eliciting responses from modern populations in Europe, Asia, and North America. Moreover, a photographic survey and mapping of 40 white and 40 black women who made *People* magazine's 100 most beautiful people list during a ten-year period found that the two groups shared similarities in conformation of the lower part of the face in relation to the neck, the lower lip projection, and the upper lip proportion. This led the authors to conclude that modern society has not only changed the classic concept of facial beauty but that in the multiethnic communities in affluent countries, women judged to be most beautiful now tend to have similar facial features arising from a mixture of "black" and "white" features.[59]

But today, as in the past, facial symmetry is considered to be attractive as a signal of both social acceptability and mate quality. This commonly held assumption has been tested in many experiments and in various ways: by asking groups of people to rate actual faces or, more commonly, sets of photographs of people (often those of celebrity actors and actresses) or computer-generated images with various degrees of asymmetry; by examining if natural variations in symmetry between faces correlate with perceived attractiveness; by manipulating the symmetry of individual faces; and even by letting subjects electronically "sculpt" their favorite variation of a reference image.

Studies have confirmed that attractiveness increases with heightened symmetry (and vice versa) for both male and female faces; that natural variations in symmetry correlate significantly with attractiveness; that individuals with more symmetrical faces also receive higher ratings of health and of such positive personality attributes as being sociable, intelligent, lively, self-confident, and balanced (symmetric faces acting as obvious "health certificates"); that these differences are not due to attractiveness stereotypes; and that more symmetrical faces are seen as more appealing in selecting a life partner, a preference attributed by evolutionary psychologists to symmetry being an honest indicator of the genetic quality of potential mates.

But measuring these effects in experimental studies using photographs is often imperfect. Evolution has made us into experts capable of fine discriminations when judging living, three-dimensional faces, as we are able to detect even very small deviations from perfect bilateral face symmetry.[60] Similarly, we can identify even small shifts in the sizes of key facial features. When people were shown a series of portraits that contained changes to the original sizes of the eyes, mouth, and nose within a range of ±14 percent (with 2 percent change between images) and were asked to identify the first image they considered clearly unlike the original, the majority of viewers set that threshold at 8 percent of the original value, even in some cases at 6 percent, with men identifying the eye-size shift much earlier than women.[61] With the average eye diameter of 2.5 centimeters this means discerning size shifts as small as 1.5 millimeters!

Moreover, experiments with digital face sculpting show that preferred faces differ among subjects who are attracted to specific features of the face.[62] Attractiveness is, undoubtedly, of great social consequence, and beauty is not entirely a social construct. Its judgment is rooted in evolution, and many studies have shown a high degree of agreement about what constitutes "beauty" not only within a particular culture but, to a significant extent, also between individuals from different cultures. Besides symmetry, we must also consider the averageness of faces—average faces being more attractive than obvious outliers—eye color, skin health and texture, hair and weight.[63] And a simple experiment (where people judge images of full faces and halves) found a strong positive correlation between the ratings of half (be it right or left) and of full faces. Based on this, we might conclude that bilateral symmetry is not a principal factor in assessing the beauty of human faces.[64]

Symmetry in human design has deep evolutionary roots. Acheulean hand axes (chipped-stone objects used about 1.5 million years ago) became the first instance of handmade symmetry imposed by hominins on their tools. By about half a million years ago, finished stone tools revealed even greater symmetries that appear to indicate an intentional concern for an attractive design going beyond the practicalities of the shaping process or functional needs.[65]

The operation of complex mechanical machinery depends on ball bearings; and in turn, their faultless prolonged functioning depends on perfect spherical symmetry. Strong materials like metals, high-density plastics, and concrete are used to make buried pipes that bring water and natural gas to our homes and take away waste. Crucially, they need to maintain radial symmetry (no buckling) for the duration of their—often generations-long—service. Required functions dictate the fact that radial symmetries define such everyday items as glasses, plates, cookware, pencils, many small tools, and round tables and flower pots, while bilateral symmetries dominate the design of furniture, garments, kitchen ovens, and receptacles ranging from envelopes to cardboard boxes.

Transportation machinery—the hulls of ships, bodies of cars, trucks and trains, and fuselages of airplanes—has outward bilateral symmetry but often a great deal of internal asymmetries dictated by specific requirements. All modern commercial jetliners are overwhelmingly symmetrical both externally (including wing-mounted jet engines) and internally (even the first-class cubicles on a few select airlines are arrayed symmetrically; food galleys and toilets are the only exceptions). But some train cars have asymmetric seating, and until there will be no need for any steering wheels, in fully autonomous vehicles, all cars and trucks will remain necessarily asymmetrical inside.

But ever since we began to build wooden and stone structures for ceremonial or religious use, architecture has provided the most impressive expressions of bilateral and radial symmetries. This dominance lasted, uninterrupted, for millennia, and even now only a minority of new structures are deliberately asymmetric.[66] Symmetrical designs

Rome's San Pietro (top, completed in 1626) and
Prague's Church of Our Lady before Týn (bottom, late 14th century).

began with the very first Mesopotamian ziggurats and with Egypt's pyramids and massive stone temples (large galleries in Karnak, Luxor; Abu Simbel, carved into a rock side with its four colossal statues of Ramesses II flanking the central doorway). Bilateral symmetries based on rectangular footprints define the iconic monuments of classical Greece (the Parthenon) and Rome (the Pantheon's portico), but Romans also built some notable circular (tomb of Augustus, the Pantheon) and oval structures (the Colosseum).

Monumental structures of ancient India and Southeast Asia are overwhelmingly symmetrical, built on plans ranging from circles (Buddhist stupas—the largest of them, the Anuradhapura dagoba) to complex square assemblies (the Hindu temple at Ranakpur) and rectangles (the temples of Angkor Wat). So were the great palaces and temples of dynastic China, with the symmetry of Beijing's Forbidden City's courtyards accentuated by marble staircases concentrating the focus on the entrances of successive halls.

Symmetries governed the art and architecture of the medieval Islamic world. The most famous embodiment of these practices is Alhambra in Granada, built between 1238 and 1358.[67] An outstanding example of later Islamic architecture is the Taj Mahal (its construction began in 1632), with the symmetry of its central cupola and four minarets doubling up as reflections in a long pool.

Medieval castles are the only large structures that were conceived and built as overwhelmingly asymmetric assemblies of buildings, towers, and walls, often in stunningly picturesque ways. Among the most audacious surviving examples are Burg Eltz above the Moselle, Corvin in Transylvania, and Orava in Slovakia.[68] Symmetry dominated Renaissance architecture, inspired by classic Greek and Roman examples, and it is on display in both the largest basilicas (such as San Pietro in Rome) and in many smaller structures. Among the most notable examples of the latter category are villas designed by Andrea Palladio, an Italian Renaissance architect, and churches by Francesco Borromini (San Carlo alle Quattro Fontane) and Gian Lorenzo Bernini (Sant'Andrea al Quirinale).[69] Symmetrical designs ruled the Baroque in Rome, Madrid, Vienna, and Prague, and their dominance remained intact throughout the 19th century—as can be seen in such disparate structures as the US Capitol, the Paris Opéra, and Milan's Galleria Vittorio Emanuele.

Building asymmetries became more common only after 1920—thanks, notably, to the designs of Bauhaus architects in Germany and to Frank Lloyd Wright's interwar houses.[70] But the age of extreme, audacious asymmetries is very recent, exemplified best by the sculpture-like structures of Canadian American architect Frank Gehry, which include the whale-like Guggenheim Museum in Bilbao and the Hotel Marqués de Riscal in Elciego with its flying color roofs. Other unprecedented asymmetries include the curvatures of British-Iraqi architect Zaha Hadid, perhaps most memorably displayed in Baku's Museum of Modern Art; Beijing's Galaxy Soho building with its dizzying parallel curves; and the Opus in Dubai with its fluidly gouged-out central space.[71] But mirror-image symmetry still rules, and it can be found not only in the design of new solitary skyscrapers but also in such paired buildings as the Petronas Towers in Kuala Lumpur by Cesar Pelli, the leaning tall structures of Puerta de Europa in Madrid by Philip Johnson and John Burgee, and Bahrain's World Trade Center by Shaun Killa.

Frank Gehry's Hotel Marqués de Riscal (2006).

But there is one art where symmetry has been conspicuously absent: painting. There are a few highly symmetrical famous paintings, above all Leonardo da Vinci's *Last Supper* (and Dalí's *Sacrament of the Last Supper*) and Perugino's *Christ Giving the Keys of the Kingdom to St. Peter* in the Sistine Chapel, but symmetry was never a dominant concern of any genre of painting, be it portraiture, still lifes, and landscapes, or mythical, religious, or allegorical works. None of the masters of the 16th to 18th centuries—Veronese or Velázquez, Rembrandt or Watteau—nor the Impressionists of the late 19th century (think of Monet's gardens, Pissarro's boulevards, Renoir's people, Cézanne's landscapes and fruits, Caillebotte's cityscapes) were preoccupied by symmetry. Their composition, perspectives, orientations, and framings are varied as dictated by the subjects of the paintings and the intent of their creators. And is it just a coincidence that Leonardo da Vinci's *Mona Lisa* earned its fame primarily because of the sitter's asymmetric smile? This was achieved by the master's use of the *sfumato* technique (the blending of edges between colors to create soft transitions) and is interpreted as a deliberate ploy to convey a cryptic, enigmatic message—one that remains elusive, despite generations of art historians trying to explain it.[72]

Asymmetry is also strongly evident in other great portrait paintings. Traditionally, a frontal view with a direct gaze was reserved mostly for images of Christ, while before 1600 European portrait paintings had a strong tendency to show more of the left side of the face, the side that has been traditionally associated with displaying more emotion. Albrecht Dürer's self-portrait at the age of 28 is one of a few notable exceptions: he painted himself Christ-like *en face*, while in self-portraits at ages 13, 19, and 26 he shows the right side of his face.[73] The pre-1600 left-side bias was especially noticeable in the case of female sitters, but the tendency was absent during the 17th and 18th centuries (though Vermeer's famous 1665 *Girl with a Pearl Earring* shows her left face) and it only partially returned after 1800.[74]

But is there not something more complex behind the appearance of many artifacts? Why do some obviously asymmetric designs seem so appealing—what is behind their asymmetric orderliness? To many art historians and mathematicians this mystery has a convincing solution: because these designs conform to the so-called golden ratio. Let's take a closer look at that claim.

Golden ratio: ubiquitous or imaginary?

As with so many analytical observations, the golden ratio has its origins in ancient Greece but we cannot be sure who identified it first—possibly many generations before Euclid of Alexandria defined it in the sixth book of his *Elements*: "A line is divided in extreme and mean ratio" if **a** + **b** is to **a** as **a** is to **b**.[75] In plain English, divide a line in such a way that the ratio between the line and the longer of the two pieces will be the same as the ratio between the longer and the shorter piece. Restated in a more rigorous way, a line segment of length 1 is divided in two pieces whereby $(1 / x) = x / (1 - x)$ or $(x^2 + x) - 1 = 0$.

Solving this equation means that x equals approximately 0.618033988. Approximately because this is an irrational number (it cannot be expressed as a ratio of two integers) with an infinite number of digits. As Chris Budd, a professor of applied mathematics, rightly concluded: "It is a lovely paradox that the most interesting thing about the golden ratio is that it isn't a ratio."[76] Rounding it to five decimals for common usage will do, and that means that the inverse value $(1 / x)$ is equal to 1.61803, and this quantity is usually denoted by the Greek letter *phi*, Φ. This is just the starting point for an intriguing mathematical journey through the supposedly Φ-rich world.[77]

A golden rectangle is easy to construct. Just draw a square, divide it in half by a vertical line, draw a diagonal line from the lower left to the upper right corner of the newly created rectangle on the right, and then pivot that line until it coincides with the base of the

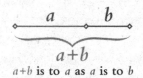

$a+b$ is to a as a is to b

Dividing a line with golden ratio.

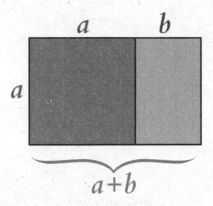

Simple to make—but how common is it really?

horizontal line protracted from the square: that point will set the length of a golden rectangle, which you will complete by protracting the top line of the original square—and when you measure the sides of the rectangle, their ratio will be golden.

The golden triangle has two sides of equal length, and this length divided by the triangle's base yields the golden ratio. Other geometric figures can have golden proportions: the rhombus, ellipse, and pentagon, and there are golden polyhedra (icosahedron, dodecahedron) and a golden pyramid. But when you start with a large golden rectangle and keep on subdividing it by golden sections, and then inscribe quarter-circle arcs into squares cut off from successive golden rectangles, you will not get a golden spiral, just a sequence of circular arcs and the curvature of the spiral changes as you go from one arc to another.

This sequence is a very close approximation of a logarithmic spiral: the resemblance of these two spirals is such that if you were to overlay their smaller-size drawings you would notice the difference between the two curves only in the largest segments. The best-known demonstration of a logarithmic spiral in nature is the shell of *Nautilus pompilius* (chambered nautilus, a mollusk living in the Pacific and Indian oceans), but it can also be found in curving animal horns and in the tails of seahorses.[78] As D'Arcy Wentworth Thompson explained in his long work on growth, "the shell, like the creature within it, grows in size but does not change its shape; and the existence of this constant

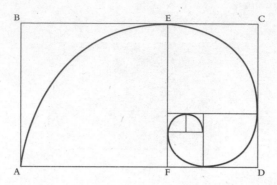

Golden subdivisions of golden rectangles do not yield a golden spiral.

relativity of growth, or constant similarity of form, is of the essence, and may be made the basis of a definition, of the equiangular spiral."[79]

When searching for evidence of the golden ratio, it is best to start with simple, indisputable facts. A rectangle is the most common shape in the designed environment (rooms, doors, desks, windows, screens, books, food packaged in boxes), and hence it should be instructive to see if I am surrounded by its golden specimens that are claimed to have an unusual appeal and that the designers are encouraged to use to create pleasing overall dimensions, get the right emotional response, or push the viewers' attention to specific areas.

If all of this were true, then I live in a very poorly designed environment: rectangular artifacts around me do not conform to the golden ratio even with ±2 percent allowance (that is, ratios between 1.58 and 1.65). They come quite a bit below it—my iPad screen (280.6 × 214.9 millimeters) has a ratio of just 1.305; a Hiroshige *ukiyo-e* (small-size horizontal) print (250 × 180 millimeters) of a Japanese landscape is 1.388; the A4 sheet of paper (297 × 210 millimeters) I feed into my printer is 1.414; the Penguin paperback of my latest book (198 × 129 millimeters) is 1. 535—or a bit above it: the screen of the Dell laptop on which this book was written (381.89 × 214.81 millimeters) is 1.779; and my old Texas Instruments Galaxy Solar calculator (152 × 86 millimeters) that I use many times a day has a ratio nearly as high at 1.767. Only when I pull out a credit card from my wallet (86 × 54 millimeters) do I get a ratio of 1.592, less than 2 percent smaller than the golden value of 1.61803.[80]

Searching for evidence beyond the confines of a room during the third decade of the 21st century means turning to the World Wide Web, but when checking the entries referring to the golden ratio it takes no time to find highly contradictory conclusions. There are lengthy (and also well-illustrated) contributions explaining "How to use the golden ratio to create gorgeous graphic designs"—and there are postings and websites dismissing it as "total nonsense" and calling it "design's biggest myth."[81] Searching further brings only more contradictions. On one side, many writers extoll the ratio's universal presence and tremendous reputation and see this enduring concept applied to works of art ranging from the Parthenon to many well-known paintings.

The painters whose works are most often mentioned as adhering to the golden ratio's rules include: Sandro Botticelli (*Birth of Venus*), Leonardo da Vinci (*The Last Supper, Annunciation*), Raphael (*Madonna of the Goldfinch*), Georges Seurat (*Bathers at Asnières, Bridge at Courbevoie*), and Salvador Dalí (*The Sacrament of the Last Supper*).[82] On the other side, there are not only dismissive designers calling it an urban legend, but also mathematicians who find virtually no traces of the proportion wherever they look.

George Markowsky (who worked at IBM's Thomas J. Watson Research Center) assembled his arguments against the ubiquity of the golden ratio as a list of misconceptions: it was not used to design the Great Pyramid or Parthenon or the UN building in New York; it cannot be found in old paintings; the golden rectangle is not the most aesthetically pleasing of innumerable rectangular proportions; and proportions of the human body have nothing to do with it.[83] Keith Devlin from Stanford started with his own experience ("I personally did not find the golden rectangle the most pleasing among all rectangles") and concluded that the enduring fascination with the golden ratio is best explained by the fact that "people, at least many people, seem to really WANT there to be numbers with mystical properties."[84]

If all of these dismissals are correct, is this just wishful thinking that has made the measure relevant? The ratio was appreciated and used by Arab and medieval European mathematicians, and it got its first catchy label (but no recommendation for widespread use) in 1509, when Luca Pacioli published his book *Divina proportione* (illustrated by Leonardo da Vinci)[85]—but "golden ratio," "golden section," and "golden number" date only to the 19th century. The term was used in 1835 in a book by Martin Ohm (his brother Georg Simon Ohm is much better known, as the unit of electrical resistance bears his name).[86]

Two decades later, Adolf Zeising, a German psychologist, did more than publish the first systematic "exposition of a new theory of the proportions of the human body"—he not only elevated the golden ratio (or cut: *goldener Schnitt*) to a universal law describing beauty and completeness in both nature and art, but presented a philosophical foundation for his claims. How excessive his conclusions were is best seen by noting that he saw the golden rule permeating "as a paramount spiritual ideal, all structures, forms and proportions, whether

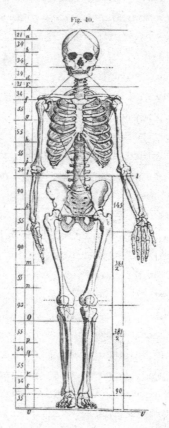

Adolf Zeising's "golden" skeleton from his 1854
Neue Lehre von den Proportionen des menschlichen Körpers.

cosmic or individual, organic or inorganic, acoustic or optical."[87] Instead of being met with profound skepticism (how could a single irrational number be related to so many phenomena?), the notion about the importance of the golden measure began to spread—albeit in more attenuated versions than Zeising's universal sweep.

In 1865, Theodor Fechner used the ratio to examine well-known paintings of Madonnas and coronations, and although he found many departures from Zeising's claim, he was sufficiently impressed to conclude that: "As much as I believe that Zeising's discovery of the importance of the golden cut must have limited application, I still consider it, even with these limits, a very interesting one, even the first one that can be described as a real discovery in aesthetics . . . and certainly it was more ingenious to present it as he did, than to limit it."[88] But the many exceptions documented by Fechner could not prevent others from searching for the golden ratio wherever possible. Since the late 19th century, the interest in applying the golden ratio to creatures and objects around us has fluctuated but persisted, and in the judgment of many of its present-day admirers it has been returned to the status to which it was elevated by Zeising.[89]

Here is perhaps the best recent example of this elevated treatment, published in 2020 by four Turkish physicians in the *International Journal of Cardiology*. After studying 162 healthy subjects, they found that two ratios of four time intervals used to assess heart functions are very close to the perfect golden ratio. But the conclusion they made regarding this very close match is astounding (no less so than the decision to publish it): "Someone may think that this is a mystical or spiritual approach, but as a general perception, our souls reside in our hearts. So it is not surprising for our hearts to beat in a Golden Ratio or namely God ratio."[90] This is an indefensibly extreme case of a highly inappropriate infatuation, but inquiries published in scientific periodicals have tried to find the ratio in places ranging from the shape of sperm heads to the crack pattern of reinforced concrete structures, with Byzantine art, pleasing smiles, and special relativity thrown in.[91]

Christopher Green, a Canadian psychologist, offered a thoughtful judgment regarding the quest for proving and disproving the appearance of the golden ratio. He concluded:

that the traditional aesthetic effects of the golden section may well be real, but that if they are, they are fragile as well. Repeated efforts to show them to be illusory have, in many instances, been followed up by efforts that have restored them . . . In the final analysis, it may simply be that the psychological instruments we are forced to use in studying the effects of the golden section are just too crude ever to satisfy the skeptic (or the advocate, for that matter), that there really is something there.[92]

In contrast, repeated analyses of human faces have not found any strong preferences for golden-section-governed beauty. Superimposing different rectangles on human faces (or differently divided segments of entire bodies) you are bound to find all kinds of ratios, and many of them will fall between 1 and 2—some inevitably close to 1.6. Add to this the likely measurement errors, and it is not surprising that many anthropometric studies searching for the ratio in human faces have ended up with underwhelming results. Papers published on this topic clearly smash the illusion, and here are just four among the latest crop. Among nearly 300 randomly selected university students in Malaysia, the mean facial indices (face length divided by face breadth) of Malaysian Indians, Chinese, and Malays were, respectively, 1.59, 1.57, and 1.54, with no significant difference among the groups. Only about 17 percent of all facial proportions conformed to the golden ratio, and the majority of students with short faces were, as a questionnaire showed, generally satisfied with their appearance.[93] Similarly, among more than 200 subjects in Turkey, only 8 percent of female and 11 percent of male faces conformed to the golden ratio— with, respectively, 92 percent and 81 percent having short faces; and three-dimensional facial analysis of Miss Korea pageant contestants and a sample of women from the general population found some notable differences between the two groups—the contestants had longer faces and smaller lower lips and chins—but did not find any significant golden ratio matches in either group.[94]

Perhaps the most convincing refutation of any golden ratio prevalence as a marker of facial beauty has come from an analysis of winners of Miss Universe and Miss Universe Thailand between the years 2001 and 2015, done by a group that is in the business of beauti-fying faces—plastic and reconstructive surgeons in Bangkok.[95] They chose more than two dozen points on a face to calculate vertical and horizontal ratios and compared them with the neoclassical canons of beauty and with golden ratios, for which they accepted a 2 percent error (1.58–1.65). Their survey and mapping of faces analyses showed that, compared with canonical beauty preferences, these beauty queens had longer foreheads, thinner lower-upper lip height propor-tion, wider nose width/corner-of-the-eye width proportion, and that their facial golden ratios were statistically significantly invalid. The greatest departure from the golden ratio (a mere 1.068) among the Miss Universe winners was for the horizontal interdacryon/interalar ratio (the distance between the eyes/width of the nose).

As for the application of the golden ratio in visual arts, Roger Herz-Fischler (before his retirement, a professor of mathematics at Carlton University in Ottawa), a diligent demystifier of the ratio, concluded that, unless documentary evidence is available, measure-ments alone are insufficient to state that an artist relied on the golden ratio to create a canvas.[96] There is nothing esthetically superior about 1.61803 compared to a simple 5/8 ratio (0.625, the inverse value of 1.6)—or if one wishes to have another irrational number why not choose the $2/\pi$ ratio 0.6366198 . . . (its rounded inverse being 1.57)?

The predilection for seeing golden ratios in art and in everyday life may be explained by the human preference for rule-based order, which in this case is made even more appealing and more explicable by deploy-ing a surprisingly simple quantitative solution. Of course, the absence of written testimonials regarding the ratio's use, and the fact of the golden ratio's closeness to other possible fractions, makes it impossible to claim (as so many writers infatuated by the notion do) that it has been used deliberately, and the only alternative is to argue that it reoccurs unwittingly because of its intrinsic superiority. When we contrast the history of the ratio's rise to popularity and the continued efforts to keep its importance alive with rigorous examination of its presence in archi-tecture, visual arts, and in past and contemporary product design, the conclusion is clear: only the most biased interpretation of this evidence would confirm the ratio's ubiquity and importance.

Still, the infatuation persists, and one of its latest demonstrations (completed in 2018) is also perhaps the most expensive one. The government of Dubai spent 230 million dirhams (more than $60 million) to build the Dubai Frame, whose "rectangular design was inspired by the 'golden ratio' of 1.618, which many architects and artists believe forms an ideal structural balance."[97] The empty frame is made of reinforced concrete, glass, and gold-colored steel, and it measures, according to the Dubai Frame website, 150 × 93 meters—producing the ratio of 1.612 (close enough). The Visit Dubai Frame website (where you can purchase tickets to ride to the top) puts its size at 150.24 × 95.53 meters (a 1.573 ratio, some 3 percent off). Whatever the actual dimensions, the structure is a hideous demonstration of having money to waste.

The overly enthusiastic search for golden ratios behind every façade, painting, or small object is clearly a misplaced quest, but there can be no doubt about the importance of properly sized design on any scale. And achieving that is not bound by adhering to a ratio that is not a ratio!

The Dubai Frame: a golden golden-ratio eyesore.

IV   SIZE DESIGNS: THE GOOD, THE BAD, THE OUTRAGEOUS

While the search for perfect symmetries and beautiful asymmetries (ruled by the golden ratio or otherwise) may often be just in the eyes of beholders, there is absolutely no dispute about the need to design our clothes and our immediate surroundings with as much attention to comfort and functionality as possible—or at least for them to be affordable. This should always have been the case, but even a slight familiarity with the history of clothing and interior and industrial design makes it clear that inconvenient fashions (from hoop skirts to tall top hats, often worn merely to uphold a wearer's status) and disregard for function—guided by attempts at self-expression and to impress others—have frequently been given priority over comfort and efficiency.[1] Unfortunately, lots of these dysfunctional designs abound.

In terms of global numbers, one of the most common (the total in use is now in the billions) individual possessions are thin rubber flip-flops with a toe thong. Formerly the footwear of the poor in warm climates (the design goes back to the ancient Middle East), now it is de rigueur on the streets of the richest western cities and on the feet of summertime intercontinental business-class travelers.[2] Too often, people wear flip-flop sizes that are obviously too small or too large, but there is really no well-fitting size for these "shoes" because they do not encase the foot and do not have a proper sole, they provide neither good lateral support nor basic vertical stability, and their wearing requires maintaining additional pressure in unnatural ways. Prolonged use can cause hammertoes due to the imbalance of muscle exertions, as well as heel pain, plantar fasciitis (inflammation of the tissue running across the foot's bottom and connecting heel bone to toes), fallen arches, flat feet, ankle and knee injuries, and knee osteo-arthritis.[3] In addition, wearing them is dangerous: they fall off easily, make people prone to tripping or being sucked into escalators, and they can become wedged under gas or brake pedals (a surprisingly common occurrence).

Furniture design is another area that has too often been guided by fashion rather than by the considerations of human anatomy and function, and it has become of even greater importance in the postindustrial world, where most jobs are indoors and require prolonged sitting. Save for the steadily diminishing share of the rich world's population that grows food, catches fish, cuts wood, mines minerals and builds structures, most of us spend most of our working time sitting indoors—and that is why ergonomics, the study of bodies fitting these designed environments, should be paramount in shaping our surroundings.[4]

And sitting now also dominates the time we travel. Pilgrimages on foot have never gone out of fashion: in 2019, the pilgrim's office in Santiago de Compostela, the destination of Europe's most popular pilgrimage, issued about 350,000 certificates to those who reached the city, starting from different points and spending different lengths of time walking.[5] But this must be seen in perspective: in 2019, the continent's three busiest airports (Paris, Amsterdam, and Frankfurt) together handled about 600,000 passengers every day.[6] Surveys show that in the USA, as well as in Japan, only very small fractions of people walk to work, while commuting for more than an hour every day is common—although in Japan it requires frequently standing up in packed morning suburban trains and subways.[7]

And there is no need to consult any surveys to know that the time people spend at home is now increasingly in front of electronic screens. That is why seats are the most obvious design challenge to consider—and perhaps the most contentious example is the economy-class seat of commercial flights. Given the sedentary nature of modern civilization, seats are worthy of a substantial detour in this chapter, but so is a more general examination of sizes deliberately changed by designers.

The premodern world was dominated by natural sizes, and the notable exception of monumental, symbolic architecture aside, the designs of our housing and their interiors (seats, tables) and of simple tools and implements used in farming (from rakes to ploughs—all overwhelmingly wooden, as metals were expensive and were smelted and deployed in limited quantities) were kept predominantly to human scale or to fit the bodies and capabilities of domesticated animals (their harnesses, saddles, implements, and the vehicles they pulled).[8] In contrast, modern developments can be seen as deliberate, serial quests to design ever-larger, even extreme, sizes—no matter the cost or inconvenience—with new records set for the height of buildings, lengths of bridges and tunnels, and dimensions and masses of industrial and transportation machines as well as of numerous consumer products.

But during the 20th century these design efforts have also gone in the opposite direction. The miniaturization of many items opened the way to further growth of other sizes and created new (and both enormous and enormously profitable) economic opportunities and consumer habits. We have been extraordinarily successful in pushing growth in both directions, but the obvious natural limits of size (from the height of a tree to the size of the largest carnivores) have their inevitable counterparts in the limits of our designs.

Human scale: ergonomics and airline seats

"Human scale" refers to sizes and the proportion of designed elements and environments—objects of daily use, furniture, interiors, buildings, neighborhoods, cityscapes—in relation not only to human physical dimensions (mass, height) and ensuing kinetic functions (reach, gait, speed of motion) but also to sensory or mental capabilities. We want to be stimulated but not overwhelmed by a variety of sizes, shapes, and colors; we have limited toleration for noise, be it excessively loud or of constant, lower intensity; and our range of optimal temperatures while working or resting is relatively limited.

Design imperatives for the objects of frequent personal use (tools, cookware, utensils, electronic gadgets) are straightforward: all of these should always be scaled to human capabilities, with good manual fit and with easy and reliable functions. Acheulean hand axes, whose elegant symmetries were noted in the previous chapter, are outstanding examples of meeting these requirements in the early

Progress in wireless communication: Douglas Engelbart's original 1963 wooden
mouse with a button and a 2020 Logitech mouse used to write this book.

stages of hominin evolution.[9] The mouse for your computer is one
of the most common examples of such applications: just compare the
first mouse, designed by Douglas Engelbart and William English in
1963—a small wooden block (7 × 8 × 10 centimeters with rounded
top edges and a single red button) riding on two thin wheels—with
the latest ergonomic shapes, which are slightly tapered and have gentle side and top curvatures.[10]

For interiors it gets more complicated. While sizes and shapes
should be kept small enough to allow for easy access, use, and to convey intimacy and harmony, comfort requires the right proportions.
Although there are many examples of modern ostentatious house
design with living rooms having the height of public spaces, such
soaring ceilings and oversize dimensions are as antithetical to human
scale as are the cramped spaces of tenements and slums, be they in Rio
de Janeiro, Lagos, or Kolkata. And there are two important corollar-
ies. Natural light should be maximized, and natural materials (wood,
stone) or steel and glass are preferable. Compared to synthetics they
cost less to produce and they are either readily recyclable (steel, glass)
or do not present disposal problems (wood eventually decays; stones

do not pollute). For buildings and neighborhoods, the challenges are still greater. Perhaps the easiest precept is to avoid any monumental scale—no oversize spaces whose emptiness dwarfs people and requires them to take very long walks between buildings—and not to design cities as if automobiles were their principal inhabitants.[11]

Urban ergonomics should aim to enhance the quality of interactions between people and their surroundings, not put obstacles in their path: walkability is, obviously, one of the most important considerations.[12] Considering the often high population densities (resulting in the push toward verticality) and the dictates of modern mass-scale transportation, this is an inherently difficult thing to do. Most readers of this book will live in cities—in 2021, urbanization rates were 75 percent in the EU, 82 percent in North America, and 92 percent in Japan—and hence completely surrounded by designed environments.[13] On the whole, these settings have evolved away from human scale—but, inevitably, they still must respect it in many ways.

Of course, there are some natural elements in cities, but even trees and shrubs planted along streets are selected for their specific desirable properties (we do not plant tall conifers, and plane trees are commonly severely pruned every winter to keep their large crowns within desired limits), and the sizes of parks and ponds often take into account the need for anticipated walking or exercising space.[14] And although many structures have outgrown human scale (skyscrapers and multilayered freeways are universal examples of this excess), there are many elements of design that remain governed by the sizes of our bodies and their parts.

In densely populated cities, apartment buildings ignore human scale, rising higher and higher: Hong Kong's public housing is composed largely of uniformly designed flats in a variety of up to 40-story-tall blocks, often built so close to each either that an inter-building conversation is possible.[15] But human scale (and scales of furniture dependent on it) must be respected: doorways have to have a minimum width and height in order to move in beds, sofas, and tables (windows, regrettably, can be made far smaller than desirable); elevator capacities must conform to prescribed safety requirements; bedrooms must fit beds.[16]

Tool design is particularly unforgiving. Ever since we began making the first tools—now believed to be as early as 3.3 million years ago—we had to make them fit our hands as we fashioned the first simple hammerstones, choppers, and flakes, and later much larger cutting flakes and the bifacial hand axes and cleavers used to butcher animals and break their bones.[17] We had to do the same with wooden and metal tools produced by the relatively complex pre-industrial societies, as the deployment of hoes, plows, and sickles by crop-growing civilizations had to conform to human anatomy. The only departure from the adherence to human constants was the need to harness draft animals used for field work and horses used in raids and wars. This required suitable harnesses for working horses (collars and other parts whose names have become unintelligible to urbanites: hame, crupper, terret, forewale . . .) and protection for horses in warfare (including often-elaborate heavy armor).[18]

All designed objects used by people must conform to the dimensions of our bodies; they must consider the abilities of our muscles and tendons to remain in confined positions; they must respect the acuity of our sight. Traditional designs were often limited by available materials and manufacturing processes. An expensive 18th-century coach could be ornately carved and have cushions on its wooden seats, but it did not have steel springs to absorb the bumps of unpaved roads, exposing passengers to punishing jolts for hours on end. Railways improved speeds by an order of magnitude, but that only encouraged longer trips—and sitting on hard wooden benches during long intercity rides brought its own prolonged discomfort.[19]

As with every kind of human endeavor, during the 19th century the design of everyday objects became a matter of deliberate engineering guided by anthropometric studies, as the empirical understanding of human dimensions provided the foundation for the new scientific field of ergonomics. As the Greek name implies (ἔργον = work), it was originally concerned with improving and optimizing labor tasks—and, rather than the usual German or French suspects, it was a Polish scientist, Wojciech Jastrzębowski, who introduced the term and wrote the first book on the topic of "science of work based on knowledge derived from the understanding of nature" in 1857.[20] During the next century, this knowledge was used mostly to optimize labor tasks and to increase productivity, but higher disposable incomes and the rise of consumer societies led to a greater attention on designs beyond the workplace—aimed at greater comfort at home and while traveling.[21]

Ergonomic principles based on anthropometric foundations are now used as much in the design of workplace environments (be they cabins of large tractors or cockpits of jetliners) as they are applied in the designs of consumer electronics, furniture, and mountain bikes. As previously mentioned, because by far the largest share of employees now work in jobs that require prolonged sitting, and because of the extensive use of electronic devices at home, ours is the most sedentary population in human history. Unsurprisingly, according to the Centers for Disease Control and Prevention more than a third of people who sit for eight or more hours a day are likely to develop one of the various musculoskeletal disorders.[22]

Designing good seats is thus more important than ever, but this quest is not without its problems and limits. Uncomfortable seats did not disappear when premodern wooden benches or stools were replaced by an enormous variety of upholstered chairs and molded plastic seats—and nothing better illustrates the challenges of deploying comfortable seats on a mass scale than a closer look at airline seating. The key sizes guiding seat design are seat, sitting, and elbow height, buttock-knee length, and hip, elbow, and shoulder breadth: make one of these too small or too large and prolonged sitting turns into an uncomfortable endurance test.[23]

Having badly designed seats at home can be remedied by standing up, walking, or choosing another seat, and these options are (within some limits) possible on many train journeys. Not so with modern, mass-scale economy-class flying. Decades ago, before airline deregulation and before the rise of mass tourism, capacity factors—especially on many long-distance flights—were much lower (in the US, flights were only half full during the early 1970s, compared to more than 80 percent full worldwide immediately before the pandemic).[24]

Moreover, seat design on long-distance flights is not only about comfort: in susceptible individuals like the elderly, hypertensive, and those with cardiovascular and digestive problems, hours-long immobility can lead to the development of deep-vein thrombosis or pulmonary embolism (the condition called economy-class syndrome) and to gastro-intestinal upsets.[25] Consequently, seats—especially on flights lasting more than six hours (the longest scheduled direct connection, between New York and Singapore, takes 18 hours and 40 minutes)—should be as spacious and as comfortable as possible, but this goal clashes with the profit imperative in a highly competitive mode of mass-scale transportation dependent on rising numbers of passengers. Before COVID-19, the annual total of passengers and revenue passenger-kilometers reached record levels after rising steadily for a decade.[26]

An economy airline seat is thus a perfect example of impossible-to-reconcile size demands (make it larger, make it smaller!), and the cause of mass-scale discomfort. The minimal dimensions are obvious: the seat must be sufficiently high above the floor to accommodate the legs bent at the right angle, it must be wide enough to fit an adult, and its pitch (the distance between the seats) must allow, as the barest minimum, space for access and for sitting with bent legs. And given the fact that flying has become a mass-scale experience, these spaces should be able to accommodate at least 95 of every 100 flyers.

Only the height of the seat off the floor is uncontroversial. The standard elevation (46 centimeters, the same as ordinary chairs) is fine for all but exceptionally tall people. Width, legroom, and seat pitch are different stories. Typical widths of house chairs are 43–46 centimeters. The width of 46 centimeters became the minimum standard with the first wide-bodied jetliners, then it increased a bit, and before the deregulation (starting in 1978) no American airline had widths less than 48 centimeters.[27] But the new era of more competitive mass flying has narrowed the width to 43 centimeters (for short-haul Airbuses and the Boeing 737s) and to as little as 42 centimeters (on American Airlines Boeing 737 MAX).[28] A recent Dutch anthropometric study demonstrated that even a seat width of 44.7 centimeters can accommodate less than 10 percent of Dutch men but about half of men in Thailand.[29]

But the main source of dissatisfaction has been the shrinking of the average pitch—the space between a point on one seat and the same point on the seat in front of it—commonly referred to as legroom. This ranges mostly between 81.3 and 96.5 centimeters (32–38 inches),

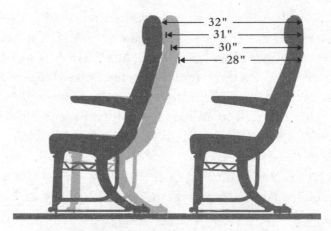

32" JetBlue (A321, E190), Virgin America (whole fleet)

31" Alaska (737-700), Southwest (737-800)

30" American, Delta, United (most domestic planes)

28" Frontier, Spirit (whole fleet)

Minimum airline seat pitches offered by American carriers.

but the average pitch on short-haul domestic flights in the US is just 31 inches (78.7 centimeters) and the minimum is now 28 inches (71 centimeters) in Airbus planes flown by airlines including America's Frontier and Spirit, Spanish Iberia, and Portuguese TAP Air.[30] And this distance shrinks even further when seats are reclined, although recent years have seen a decline of the recline, to just 2 inches, or to "pre-reclined" seats that cannot lean back at all.

Problems with airline seating have naturally been made worse by the post-1970 rise in obesity. Among the major nations, the United States has been the leader of this shift: the shares of overweight people, with a body mass index of 25–30, and obese people (BMI of more than 30) reached nearly 36 percent—and by 2018 it rose to 42.4 percent, the first-ever instance of the national rate passing the 40 percent mark.[31] Adding about 31 percent of the population that is overweight means that by 2020 almost three out of every four Americans did not fit within the healthy weight range as defined by the Centers for Disease Control and Prevention.

Moreover, in too many cases this shift in body size has reached morbid levels of obesity, with the BMI above 40. By 2020 more than

15 million adult Americans were morbidly obese, and this has necessitated new designs for beds and chairs as well as for "bariatric"
stretchers, ambulances, and mechanical lifts.[32] The lifts are needed to
move such people not only from their home beds to hospital beds but
to assist in transferring them from their beds to specially designed
wheelchairs, toilets, and baths. Heavy-duty patient lifts can now
handle bodies between 500 and 1,000 pounds (up to 454 kilograms).[33]
And the trend toward excessive body sizes has been nearly universal.
The WHO's latest estimate put the total of overweight adults at
nearly 2 billion (almost 40 percent of the total) and obese adults at 650
million, and noted the increasing frequency of excessive weight
among teenagers and children.[34]

Leaving the extremes aside, increased frequency of excessive weight
makes almost all normally sized airline seating inadequate.[35] Business-
class seating is better and as of 2022 there were still 20 airlines that
offered even more space in first class: none in North America, but Air
France, British Airways, Lufthansa, and Swiss in Europe, Qantas, 10
airlines in Asia, and 5 in the Middle East (with some airlines having
even fully enclosed pods). All an average passenger can do in order to
get the most spacious economy seat is to check out SeatGuru, which
has plans, dimensions, and seat reviews for 800 airlines.[36]

There do not seem to be any improvements in the offing: in 2017,
Flyers Rights, the passenger rights group, petitioned the US Federal
Aviation Administration (FAA) to set the rules regarding the minimum pitch, but no regulations came into force before the pandemic—
making such efforts moot.[37] And the tightening pitch is not the end of
it: in 2013, French inventor Bernard Guering filed a patent, on behalf
of Airbus, for "a seating device with reduced bulk"—a fancy term for
a row of smallish bicycle seats with low backrests mounted on a tube.[38]
Even worse, in 2010 Ryanair had plans to introduce "vertical seats,"
with crammed passengers strapped to thin, near-vertical (an incline of
just a few degrees) beds.[39] Ryanair even claimed that 2 out of 10 polled
passengers said they would fly strapped upright if the fare was halved!
And there are also designs aimed at packing in more people by placing
seats on top of existing seats (double-decker seating).

Airline aficionados might point out some good news—above all a

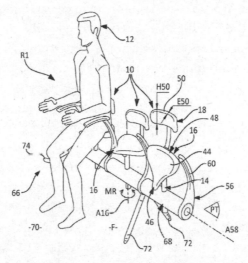

Illustration from the US patent application filed by Airbus in 2014 for
"Seating device comprising a forward-foldable backrest."

new economy seat prototype for short- to mid-haul flights developed
for Airbus by Layer, a British design company, using a single-piece
sling seat fastened to a lightweight perforated composite frame. But,
as with so many prototypes, and with so many announcements of
digital wizardry, we will have to wait to see if such seats actually get
installed in at least hundreds if not thousands of jetliners—and, most
importantly, what pitch will be chosen for their commercial use.[40]

Changing sizes: incomes, machines, and vanities

So far I have explored some notable examples of designed objects and
looked at how to achieve better results by deploying ergonomics.
These outcomes can be commercialized rapidly and—given the
electronic design and rapid modern prototyping (some using 3-D
printing)—new mountain bikes or new airplane seats can be dreamt
up and on the market in a matter of months. That is all fairly straightforward. But it is not always that easy. Here I will look at changing
sizes in more complex designs.

Even as recently as 1800 the world was overwhelmingly rural. In most Asian and African societies more than 90 percent of people lived in villages; in the US the share was about 90 percent; in Europe the UK was a major exception, with some 33 percent of the population being urban.[41] And, much like their premodern counterparts, early modern populations and economies changed very slowly—the sizes of villages and towns remained stagnant or grew almost imperceptibly because high birth rates were negated by high death rates, and because crop yields could be hit by unpredictable downturns. Reconstructions of long-term population growth show that typical annual increases remained very low (less than 0.2 percent until the 18th century) and that recurrent spells of food shortages, including famines, were not eliminated even in Europe and Japan until the 19th century. Europe's last famine was in Ireland between 1845 and 1852; Japan's last famine lasted from 1833 to 1837.[42]

Similarly, per capita economic growth was slow to nonexistent, and for most people their life's horizons were tightly circumscribed due to the absence of affordable long-distance land transportation. This combination of low incomes and poor transportation capabilities created the world of barely changing sizes and repetitive practices. Traditional staple crop harvests (wheat yields just around half a ton per hectare; rice less than a ton), the ways the crops were produced (relying on harnessed draft animals, hand-seeding, scythes and sickles, carts and flails), and the personal possessions of peasant families (a change of clothes, shared beds, few kitchen items, benches rather than chairs) remained as predictably unchanging and meager as opportunities in education.[43]

The best reconstructions of Roman per capita income during the early empire (second century BCE) point to less than 400 *sestertii*, which would be about $1,000 in 2020 monies, and later reconstructions show that per capita income was no higher a millennium later in England, or by 1500 in Ming-dynasty China.[44] Economic inertia is perhaps best illustrated by long-term per capita growth rates of incomes: a near-stationary 0.03 percent per year for Spain between 1270 and 1850; none for central and northern Italy between 1300 and 1913; and −0.1 percent for China between 1020 and 1850.[45] Inertia, familiarity, and predictability ruled.

There were always individuals, small groups of adventurers, or armies of conquest that had ventured further afield, but most lives unfolded within territories whose sizes were circumscribed by walking and riding speeds, or by the pace of horse- and ox-drawn carts and carriages (walking averaging about 5 kilometers per hour, although a fast messenger on horseback could cover more than 150 kilometers a day, but ox carts on poor roads might do only 15–20 kilometers daily).[46]

But even during the early modern era, some sizes were enlarged significantly. Most obviously, European journeys of discovery—initiated by the Portuguese and the Spanish during the late 15th century, and later carried on with major Dutch, English, British, and French endeavors—had steadily expanded the size of the known and mapped world: by 1800 only the farthest Arctic and Antarctica remained *terrae incognitae*. But only the 19th century, and especially its last four decades, became the time of unprecedented, epoch-making change, as the prolonged stasis and slow progress of the premodern era were replaced by developments driven by scientific and engineering advances that had rapidly created new economies, new societies, new accomplishments, and new expectations.[47]

Despite the setbacks and horrors of the two world wars and scores of other inter- and intra-national violent conflicts, this unprecedented rate and breadth of change continued during the 20th century, with results ranging from admirable to unwelcome. Many detailed writings review the enormous variety of innovation-driven post-1860 designs of technical artifacts and production sequences, ranging from energy converters (engines, turbines) and heavy industrial processes (metallurgy, the bulk production of chemicals, the mechanization of factory tasks) to the rise of new mass-volume enterprises (chemical syntheses, drug development, electric appliances) and the adoption of new transportation machinery and its requisite infrastructure (road, rails, bridges, tunnels).[48]

Most people do not realize how extraordinarily large these gains in size of machines and processes have been. Perhaps the best way to illustrate these advances is to focus on a fundamental determinant of modern civilization: the power of prime movers (natural or mechanical sources of kinetic energy) and the consequences of their applications in food production, industries, transportation, and households. Preindustrial societies were energized primarily by animate prime movers: by people of all ages (child labor was common) and by domesticated animals harnessed for field work or transportation. Waterwheels were the most common inanimate energy converters.[49]

By the end of the 18th century their best designs remained small, with power equivalent to no more than 16 horses; by the middle of the 19th century waterwheels (up to five times more powerful) became the most important inanimate prime mover, delivering more than twice as much useful energy as small steam engines. Steam engines dominated industrial and transportation needs until after the First World War, and subsequently three types of inanimate prime movers took over all high-power applications: internal combustion engines, steam turbines, and gas turbines.

No other fuel converter is now more common than a gasoline-fuelled engine, and even in small cars it can now deliver more than 100 kilowatts—the equivalent of more than 130 horses (1 horsepower = 745.7 watts)—while in the US, cars now average 135–150 kilowatts and SUVs and pickup trucks, the dominant choices on the US market, have engines mostly between 200 and 250 kilowatts (different models of the Ford F-150, the bestselling US vehicle, rate between 220 and 335 kilowatts).[50] This means that more than a billion people (the global count of all road vehicles is now approaching 1.5 billion) are now individually commanding machines whose unit power is commonly an order of magnitude higher than the power of the largest mid-19th-century industrial waterwheel designs used in large flour mills and textile factories.

But a more proper comparison should use today's largest energy converters deployed for commercial uses in transportation and industry. In transportation, the two largest internal combustion engine sizes are the two-stroke diesel engines powering large ocean-going ships (oil tankers; bulk carriers loaded with ores, cement, grain; and container ships mostly carrying manufactured products in steel cases) and the gas turbines (jet engines) that allow passenger airplanes to cruise at speeds approaching the speed of sound.[51] The largest marine diesels now have a power of 80–90 megawatts and, similarly, a Boeing 747 needs about 90 megawatts for takeoff (when cruising, energy needs are much lower).

This means that the captain of a large vessel or of a wide-body jet-liner controls—now through complex electronics—power that is up to 300 times that of a F-150 driver, nearly 1,000 that of a Honda Civic owner, and almost 100,000 that of an early 19th-century waterwheel used to mill flour for a small town. And the multiples are similarly large when we compare early industrial waterwheels with steam and gas turbines, which are considerably more powerful than ships and planes. The largest steam turbines used to generate electricity have a power of more than 1,000 megawatts (the record size is just over 1,700 megawatts); common sizes are 500 megawatts and more. The largest stationary gas turbines now also rate more than 500 megawatts—that is, 100,000 times the size of small early waterwheels![52]

These orders-of-magnitude size increases of prime movers have enabled us to produce a surfeit of food and plenty of affordable con-sumer items, as well as giving us access to information and increased mobility. By far the most fundamental consequence of these chang-ing sizes has been in crop production: without mechanical prime movers, most people would have had to stay on the land in order to produce sufficient food supply, precluding mass migration to cities and the rise of modern industrial production.[53]

A century and a half ago, the combined power of four strong horses harnessed to a plow or harrows (the typical way of preparing cropland for planting during the late 1800s) was no more than 3 kilo-watts and about half of the labor force was on farms. In recent decades, North America's bestselling John Deere tractors that do the same tasks have power of no less than 120 kilowatts, with only 1.3 percent of people needed to work on farms in 2020.[54] This means that the relative increase of tractive power during the past 150 years (120 / 3 = 40) was exactly the reciprocal value of the relative decrease in the share of the US agricultural labor force (1.3 / 51.3 = 0.025). The rela-tive size of farm labor shrank in inverse proportion to the relative increase of machinery power.

Size increases in material production have been no less impressive. Between 1800 and 2020 the internal volume of modern blast furnaces increased about 60-fold, and their daily output 3,000-fold, making steel (decarbonized iron) inexpensive and abundant.[55] An increase in the size of ammonia synthesis has been even more consequential: it has gone up 1,000-fold in a little over a century.[56] This has led to higher crop yields, lower food prices, and (due to the possibility of devoting more land to animal-feed cultivation) increased consumption of animal protein in dairy products, eggs, and meat.[57]

Another way to trace long-term developments is to look at adaptive changes as opposed to deliberate quests for larger or smaller products. On the individual level, the most common example of adaptive change has been the need to supply the market with larger apparel and shoe sizes, as dictated by taller and heavier human bodies. But in the case of women's clothes in the US this shift has been complicated by the countertrend of "vanity sizing," where garments are labeled as smaller than they actually are in order to make the buyers feel better about their bodies. For example, in 1958 Marilyn Monroe (bust 34 inches, waist 25 inches) would have been a size 12 (in the UK, size 16)—but in 2020 a size 12 would fit a female whose bust is 5 inches larger and whose waist is 7 inches larger.[58]

Traveling around any modern country it is obvious that the introduction and large-scale expansion of hard-top (asphalted or reinforced concrete) roads has been one of the most consequential adaptive changes brought by the production of affordable automobiles. Societies had to address the problem of having cars *before* they had roads, and to fix it by road-building programs on an unprecedented scale. When in 1908 Ford introduced its Model T, the world's first mass-produced vehicle, less than 10 percent of all US roads were surfaced, and less than 10 percent of those were of good quality.[59]

America's first short concrete highway was built in 1913 in Arkansas; for many years after the First World War, the famous Route 66 (Chicago to Santa Monica) remained mostly gravel, with asphalted stretches as narrow as 3 meters; and the country's first major all-concrete highway, the Pennsylvania Turnpike, only opened in 1940 when the country had more than 32 million road vehicles.[60] Construction of America's Interstate highway system began in 1956, and its size in 2021 was about 78,000 kilometers—less than half of China's new National Trunk Highway system of about 160,000 kilometers, which is to be expanded 50 percent by 2035.[61] Eventually, all of these growth trends will slow down, reach their limit, and some of them will retreat. This is the way of the world.

Limits of size, or why some records will remain unbroken

The study of growth is, inevitably, the study of the limits of size.[62] The universe may keep on expanding, but on Earth all growth eventually ends—and given the enormous variety of growing organisms, natural features (mountains, water bodies), and anthropogenic processes (cultivation, construction, production), there is a commensurate variety of natural and man-made limits to growth. For organisms it is the typical and maximum size potential determined by their evolution. For photosynthesizing plants, reaching their limits is determined by the availability of solar radiation, water, and all requisite macro- and micronutrients, as well as the absence of pests and of viral, fungal, and bacterial diseases. For animals and humans it depends on the adequate supply of the three indispensable macronutrients—carbohydrates, proteins, lipids—as well as of vitamins, minerals, and water, and on the absence of debilitating diseases.

What are some of the major factors that limit the size of artifacts—from simple tools to complex machines? In the most obvious physical terms, the limits of our designs are governed by the limits of the materials we use to make them—but there is no shortage of historical examples of designers going too far. *Quinqueremes*, ships with 300 oarsmen, were common in antiquity, and they were followed by larger designs culminating in the construction of *tessarakonteres* during Ptolemy IV Philopator's reign (222–204 BCE). That vessel was to carry more than 4,000 oarsmen and nearly 3,000 troops, but its weight (including heavy catapults) made it practically immovable.[63]

A similar fate met the Spruce Goose, the world's largest wooden plane—powered by eight reciprocating engines and with a wingspan of 98 meters. Howard Hughes got it airborne only once, on November 2, 1947 for about a minute.[64] And while the double-decker Airbus

The Spruce Goose (which flew only once, in 1947) and
Airbus A380 (made between 2003 and 2021).

A380, introduced in 2003, was eventually bought and flown by 15
airlines, the (always questionable, always too big in the age of direct
connections rather than hub-to-hub flying) design met its early end
during the pandemic: some airlines will keep it in service, but its production ceased after making just 251 airplanes.[65] In contrast, there
have been many record-size designs that went on to conquer their
respective markets. Nearly 1,600 Boeing 747s, planes that began the

era of affordable mass flying, were made between 1969 and 2022; the Ford F-150 became not only the bestselling pickup truck but America's bestselling vehicle; and the Big Mac has been the embodiment of America's favorite fast-food empire.[66]

But in most instances, the history of engineering could be seen as a constant quest to circumvent physical restrictions by introducing new materials and deploying them in new ways—and, in turn, these admirable measures eventually face their own inherent limits. Of course, human inventiveness has only followed the great example of evolution, which introduced new materials and new structures to meet previously impossible demands. Among its notable material innovations are the coral skeletons of calcium carbonate that create massive reefs; lignin (complex organic polymer), which makes up about a quarter of all tree trunks; feathers composed of twisted and cross-linked protein strands, which enable avian flight; and human skin, whose tear resistance, thanks to the ability of collagen fibrils to slide past each other, is (depending on the strain) as high as 15–21 megapascals—compared to just 2–5 megapascals for concrete.[67]

Examples of these evolutionary innovations and adjustments could fill a sizable book but, obviously, all of them can only go so far before other limits intervene. For example, a recent discovery shows that a Cretaceous pterosaur (genus *Alanqa*, living some 95 million years ago) had very light vertebrae, constructed in a way that considerably reduced its overall skeletal mass even as it significantly increased the buckling load and made it possible for this large flyer—with a wingspan of 4–6 meters—to pick up and carry relatively heavy prey without risking damage to its cervical spine.[68] The largest toothless pterosaur with a long, stiffened neck, *Quetzalcoatlus northropi*, had a maximum wingspan twice as large—but millions of years of Cretaceous evolution could not come up with any mechanical and structural advances that would allow even-larger flyers.[69]

Unlike those long-extinct pterosaurs, tree sizes offer an example of a readily verifiable structural limit. The wood of tree trunks is composed of cellulose (a natural polymer of glucose), hemicellulose (also a natural polymer, but one made up of carbohydrates of low molecular weight), and lignin (about a quarter of the total mass).

The highest trees—giant sequoias, Douglas firs, and *Eucalyptus regnans*—can grow to 110–125 meters: it is impossible to have a tree as tall as even a middling-size (more than 200 meters tall) Manhattan skyscraper.[70] We cannot point to a single limiting factor: both mechanical (loads on woody self-supporting stems) and hydraulic (distribution of water inside the trees) factors constrain the ultimate height, but evolution may have optimized tree survival and repro-duction of trees rather than their longevity and massive size.[71]

Long before modern studies by plant morphologists looked into tree growth limits, Galileo Galilei (I will return to his observations in the next chapter, on scaling) recognized in his 1638 *Dialogues Concerning Two New Sciences* that the tallest trees could not be taller than about 140 meters: "I am certain you both know that an oak two hun-dred cubits high would not be able to sustain its own branches if they were distributed as in a tree of ordinary size."[72]

Inevitably, these limits of evolution have their counterparts in the maxima of engineering accomplishments. Comparison of tensile strengths (the force, measured per unit of area, needed to break a material) explains why modern skyscrapers can go well above 500 meters: in megapascals the tensile strength is 400–800 for steel, but only 70–140 for wood.[73] The world's tallest building (Burj Khalifa) has already surpassed 800 meters; Jeddah Tower (whose construction was put on hold in 2018, but not because of encountering technical problems) was designed to reach 1 kilometer; and substantially taller buildings are possible when using buttressed-core foundations with a hollowed-out base, a design akin to the Eiffel Tower.[74]

Such buildings could go well above 2 kilometers but their base would have to be commensurately large. For a 4,000-meter building whose design was contemplated in Japan during the 1990s (X-Seed 4000, taller than Mount Fuji at 3,766 meters), the hollowed-out base would have covered an area of around 1,000 square kilometers (a square with a nearly 40-kilometer side—for comparison, 23 densely popu-lated wards of Tokyo that house more than 9 million people cover less than 630 square kilometers)! All too obviously, other factors than structural properties of building materials would preclude such heights.

How could such enormous foundation areas get access to natural light? How much space inside the structure would have to be taken up by yet-to-be-invented super-elevators (not only for people but for food and supplies)? Current elevator systems do not go higher than about 600 meters, so a mile-high building would require two elevator changes to get to the top. Perhaps most importantly, who would finance the purchase of the land required for such a project and its inevitably very lengthy construction? For private investors, return on the upfront capital would be dismal, but would a government step in to build these vertical equivalents of not-so-small cities? And how much would it cost to insure this megastructure—what measures would have to be taken to ensure its speedy evacuation in case of a fire or an earthquake?

These financial, social, and organizational challenges are hardly unique to the quest for super-tall buildings; they are repeatedly encountered wherever size increases are limited not by the properties of materials or by the absence of technical solutions but rather by diminishing returns on investment, increasing capital costs, and reduced flexibility. Such caps on growth are common but, because of their importance, three examples are particularly notable. The first is the ocean-borne trade of crude oil, the fossil fuel that has made it possible to raise the standard of life to unprecedented heights for an increasing share of humanity—and that will remain (contrary to recent exaggerated expectations of the imminent demise of fossil carbon) a major source of primary energy for decades to come.

Oil became so dominant because it is easy to transport and because it is so versatile. The extraordinary concentration of its deposits in the Middle East (Saudi Arabia, Iran, Iraq, Kuwait, the UAE) means that its exports must rely on large tankers: they carry about two-thirds of recent crude oil exports; the rest moves through pipelines.[75] Tanker sizes are best expressed in terms of deadweight tonnage (dwt), the total weight that the vessels are able to carry. Maximum crude oil tanker sizes increased from only about 20,000 dwt after the Second World War to 100,000 by 1959 and to more than 300,000 by the early 1970s. The expectation was eventually to have ships of 1 million dwt—but the *Seawise Giant*, launched in 1975 and enlarged later to 564,650 dwt, was as large as we have ever gone.[76]

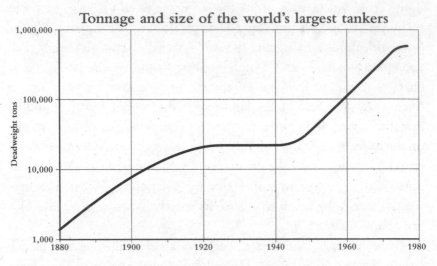

Growth of crude oil tanker capacity, 1880–1980.

So why have we never built a tanker of 750,000 or 1 million dwt? The economies of scale that drove the post–Second World War trend for increasing tanker sizes did not stop at half a million tons (the larger the ship, the lower the cost per ton of oil shipped per unit of distance), and there were no technical obstacles: the same construction methods used for a 500,000-dwt ship could be relied on to build a mega-tanker of an even larger scale, and we could deploy powerful diesel engines to propel it on journeys of thousands of kilometers without stopping. But as we shall see time and time again, steadily increasing size is accompanied by an inexorably diminishing return on investment. When Japan dominated the industry, doubling the size from 60,000 to 120,000 dwt reduced the building cost by 30 percent, but the subsequent doubling to 240,000 dwt saved only about 15 percent and any size above 300,000 dwt might be, considering the limited flexibility of use of such a large vessel, counterproductive. Let me explain why.

Bigger ships have a deeper draft (distance between the waterline and the bottom of a ship's hull), which restricts their routing and their ports of call. Only a handful of deep-water ports are deep enough to accommodate a megaship whose draft would exceed that of a

500, 000-dwt vessel, and those ships could not pass through either the Panama or the Suez Canal because, even after its enlargement, the Suez Canal allows 20.1 meters of maximum draft and ships of no more than 200,000 dwt, while so-called Panamax ships (the largest ships allowed through the Panama Canal) have to reckon with only a 12-meter draft and a maximum size of 65,000 dwt. And the Straits of Malacca route (the shortest way from the Middle East to the Far East), with a minimum depth of 27 meters, only allows tankers of up to 300,000 dwt.[77]

Obviously, gargantuan ships would have very limited flexibility both in terms of ports and routes—but they would also be riskier to operate. In the case of an emergency, they need long distances to stop and to turn around (*Seawise Giant* needed 9 kilometers from its top speed, and its turning radius was about 3 kilometers). They also cost more to insure, and this concern became particularly prominent after the world's two largest catastrophic oil spills—those of *Castillo de Bellver* in South Africa in 1983 and *Exxon Valdez* in Alaska in 1989. The latter accident cost the company as much as $7 billion in clean up costs, fines, penalties, and claim settlements.[78]

The first oil tanker was launched in 1886, and the growth of this remarkable machine ended nine decades later—replaced with vessels whose deadweight tonnage was more than 100 times larger. What happened to the world's largest tanker? In 1988, *Seawise Giant* was attacked and damaged during the Iran-Iraq War, relaunched in 1991 and renamed *Jahre Viking*, then retired, renamed *Knock Nevis*, and repurposed as a floating storage and offloading unit offshore of Qatar. In 2009 it was sold to Indian shipbreakers and renamed, for the last time, *Mont* for its final journey to Alang beach in Gujarat, where old ships go to die (that is, to be emptied and cut up for steel scrap).[79] France built four other ships nearly as large as the *Seawise Giant*—so-called Batillus-class vessels of 555,000 dwt launched between 1976 and 1979. They had even shorter lives, with three of them scrapped by 1986 and the last one by 2003.[80] Much like giant sauropods, giant tankers are gone forever.

Another useful example of the limits of size in the energy industry is that of wind turbines. These gangly triple-bladed machines are seen as one of the key ways to end modern civilization's dependence on fossil fuels, and they have been growing steadily taller in order to accommodate ever-larger rotors. This is as expected, because wind velocity increases with height above ground. If a turbine's hub is 150 meters above ground (as it is for one of the world's tallest designs, GE's Haliade-X), it might benefit from a mean annual wind speed up to 20 percent higher than a machine whose hub is at just 50 meters.[81] Moreover, turbine power goes up with the square of the radius swept by its blades, and hence the GE machine, with 107-meter blades, would (everything else being equal) generate about 7.2 times more electricity than a turbine with 40-meter blades.

Building taller wind turbines with larger rotors is the most obvious way to boost their generating capacity, but this quest runs into unavoidable problems of scaling and gravity. If the larger turbine rotors used the same materials, the same shapes, and the same relative thickness as the smaller blades, their mass (M) would have to scale cubically (to the third power: M^3): a blade twice as long would have to weigh eight times as much. This would make for unacceptably heavy rotors, but advances in structural and material design (the use of such lightweight materials as composites and balsa, the world's lightest wood) were able to keep the mass-growth exponent to a sub-cubic rate—as low as 2.1![82]

As a result, a blade twice as long will weigh only 4.3 times as much ($M^{2.1}$). GE's record-breaking design has blades 107 meters long weighing 55 tons, and with even larger sizes mass becomes a critical consideration: we will never have wind turbine blades three times as long as the GE machine. Their mass would be 10 times greater (each about 550 tons), putting enormous strain on the nacelle (the streamlined container housing the equipment converting kinetic to electric energy, to which the blades are attached) and on the turbine tower.

Limits at the opposite side of the size spectrum have been encountered, and repeatedly overcome, by placing more components on tiny silicon chips. Moore's law of 1965, anticipating rapid doublings of components placed on a microchip, held fairly well for the remainder of the 20th century.[83] The number of components grew by an additional six orders of magnitude, to more than 10 billion components in a single microprocessor.[84] This resulted in the steadily declining cost of microelectronics—and this increasing affordability

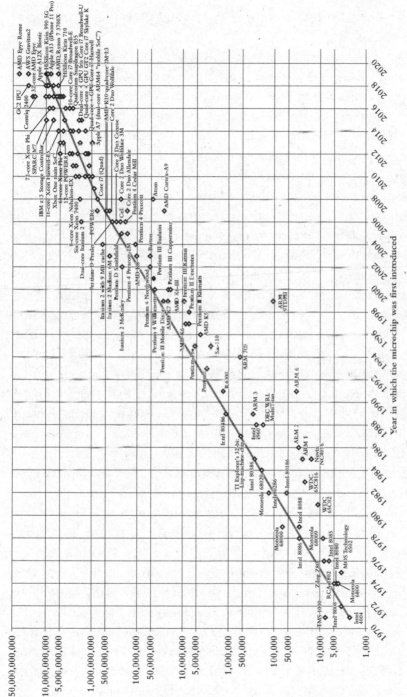

The number of components on a microchip used to double every two years, but there has been a discernible post-2008 slowdown.

means that microprocessors are now running everything from cars to phones.[85]

But it is impossible to keep on packing ever-rising numbers of components on a finite area. This "law" has already hit its limit. As long-established chip production techniques have become more difficult (and very costly) to apply—as we will see below—the future growth of computing power will have to reckon with the end of Moore's law. The first clear sign of approaching limits came in 2008, when the slowest supercomputer on the list of the world's 500 most powerful machines missed its expected operating speed increase.[86] Five years later, performance trajectory began to flatten for all the 500 top systems. The reason is that so-called Dennard scaling had plateaued. In 1974, Robert Dennard and his coworkers at IBM noted that, as transistors got smaller, it was possible to increase operating speeds without increasing overall circuit power consumption.[87]

Eventually, this advantage came to its end because transistors began overheating, and this prevented further acceleration of operating speeds. As a result, operating frequencies, after rising from 100 megahertz in 1994 to 3,000 megahertz (3 gigahertz) by 2004, settled mostly at around 3.5 gigahertz, with 5 gigahertz as the upper limit— well below the speed that would have been expected if Dennard scaling continued.[88] The other limit for computing power is related to microchip manufacturing. Optical lithography (printing with light) has been used in chip fabrication plants (fabs) to emplace ever-thinner transistors: the first transistors were 80 micrometers wide; today's smallest width is just 5 nanometers, or a size reduction of 16,000 to 1![89] Another astonishing comparison is that, at 5 nanometers, the width of a transistor is equal to about 20 silicon atoms.

Every width reduction requires enormous investments in new fabs, whose cost rose from a few million dollars to more than $10 billion apiece.[90] Even Intel, the company that pioneered this manufacturing process, gave up this extraordinarily costly race to the bottom, leaving the field to Taiwan's TSMC, South Korea's Samsung, and American GlobalFoundries. But this is not the end of that miniaturizing spiral: in 2019, GlobalFoundries announced that it will stop all development of the 7-nanometer design, leaving only the two Asian companies in the sub-10-nanometer range, but in 2021 IBM announced its 2-nanometer design that should perform 45 percent better and use 75 percent less energy than the 7-nanometer chips.[91]

In this chapter I have tried to explain the two principal challenges of modern artifact design: that we should be paying more attention to human scale and applying ergonomic understanding as widely as possible; and that, because of inherent limits, modern civilization will not be able to design its way out of its many predicaments. Encouragingly, both realities share a common attribute: the need for rational adjustments, and for staying within sensible limits. Of course, we can keep ignoring them, as we have done repeatedly in the past, or we can take them more seriously.

V   SIZE AND SCALING

How do organisms change as their body size changes? Does a woman weighing 75 kilograms have a heart and lungs that are 50 percent more massive than her 50-kilogram cousin? Is her skin area or blood volume 50 percent larger? Does a tall and heavily muscled 100-kilogram man need to eat twice as much as a small and slim 50-kilogram teenager? We want to know not only because these questions sound intriguing, but also because of the practical necessity to answer them as best we can. When a woman is burned in a serious house fire, specialists who treat her want to know what proportion of her skin has been affected—and for that they must know her body surface area. When an injured sumo wrestler faces a prolonged recuperation in hospital, he should get enough food to facilitate his recovery—and for that a dietician should know his daily food requirements. And can the heart of a small but healthy teenage girl killed in a car accident be transplanted into the body of a 50-year-old man of average weight?

Obviously, answering similar questions is no less important for all other organisms, be they the cows and sheep we keep for milking or the pigs we grow for meat and fat. Do we have to feed the largest cattle breeds 50 percent more than the animals weighing only half as much? Does a very large Danish bacon-producing pig need proportionately more feed than a smaller Canadian animal bred to pack on the leanest possible meat? Rephrased in general terms, do animal skeletons, surface body areas (skin, hide), blood volume, internal organs (brains, hearts, lungs, livers, kidneys), and key functions of the living bodies (metabolism, heartbeat, breathing) change in direct proportion to their changes in weight—or does something else happen?

You already know some of the answers. If you have ever held a tiny bird in your hand (the last time I did so was when I was warming up a hummingbird who nearly died during an unseasonably cold

September night), you know that his heart beats much faster than yours. And you know that squirrels move around almost constantly in search of food. You have seen Serengeti lions dozing off in some sparse shade after gorging on a freshly killed buffalo: they will not have to eat for a few days and can do without any meat for a week.[1] But does it mean that all smaller hearts beat faster and that all large animals can feed only intermittently? And if there are cases of clearly disproportional scaling, which organs increase faster than total body weight and which ones have relatively smaller shares of body mass as animal size increases?

And can a very small, or a very large, organism experience the world in the same way as does a body whose mass is between these extremes? Jonathan Swift, who became the dean of St. Patrick's Cathedral in Dublin in 1713, built his masterpiece *Gulliver's Travels* on the mistaken assumption that such a thing—invariable functions and behavior despite enormous, downward and upward, changes in body

Lemuel Gulliver as portrayed in the frontispiece of Swift's 1726 masterpiece.

mass—is perfectly possible. For this reason, I cannot think of a better way to dive into the fascinating subject of scaling than to turn to the adventures of that restless ship surgeon, later captain, Lemuel Gulliver of Redriff, whose sojourns in the lands of diminutive Lilliputians and gigantic Brobdingnagians were narrated by Swift in that perennial favorite, the book of travels "into Several Remote Nations of the World" published in 1726.[2]

Although my indictment of Swift's errors will be methodical and factually unassailable, I hasten to add that his scaling blunders do not detract in the least from the brilliance of his work. Samuel Johnson's often-quoted dismissal of the book ("When once you have thought of big men and little men, it is very easy to do all the rest"[3]) is not only an unjustifiable put-down of a clever artistic choice, it also betrays Johnson's ignorance of many profound implications of much increased, or much diminished, size: would his version of the tale avoid Swift's scaling errors, and would it matter? Swift's imaginary travels remain one of my favorite pieces of acerbic, unsparing, and captivating prose: I know that he wrote "to vex the world, rather than to divert it."[4]

So, with apologies to the dean, let us set out the basic facts first. The closest Swift comes to telling us Gulliver's actual height is when, in Brobdingnag—the land occupied by the giants—he describes him as "not so big as a splacnuck (an animal in that country very finely shaped, about six feet long)." Six feet is 1.82 meters. The average height of men in 18th-century England was less than that (about 170 centimeters), so let us assume that Lemuel Gulliver was 175 centimeters tall.[5] Earlier in the book we are told that in Lilliput (the island of tiny people), mathematicians took "the height of my body by the help of a quadrant" and found that it exceeded "theirs in the proportion of twelve to one." That would make them about 15 centimeters tall (the length of a fountain pen or a small toothbrush), and as the proportion worked the other way in Brobdingnag, its giants would be 21 meters tall—that is, the height of a seven-story building or a good-size spruce tree. As for body mass, Lilliput's mathematicians "concluded from the similarity of their bodies, that mine [Gulliver's] must contain at least 1724 of theirs."

In an 1803 colored etching, James Gillray made the king of
Brobdingnag (comparing Gulliver's height and the king's head)
to be of below-average height—a bit more than 15 rather than 21 meters.

Swift's errors, Galilei's explanations

The previous sentence contains two mistakes: one only a minor algebraic error, the other one a fundamentally erroneous assumption.
Swift's small error was just a wrong computation: 12^3 ($12 \times 12 \times 12$) is
not 1,724 but 1,728. Swift does not give Gulliver's weight, but from
modern anthropometric studies we know that a 1.75-meter-tall,
medium-frame adult male would weigh between 64 and 70 kilograms.[6] Taking 67 kilograms as Gulliver's weight, a Lilliputian would
then be just shy of 39 grams, no bigger than a large shrew. Conversely, a Brobdingnagian adult male, whose mass would be 1,728
times that of Gulliver's, would come to 115,776 kilograms or nearly
116 tons—larger than any of the largest-known dinosaurs but with a

fundamental difference: that mass would be contained in an upright bipedal body!

We do not know if Swift actually followed his assumptions and calculated this monstrous weight, but if he did, that enormous mass should have given him plenty of pause—or not: the assumption was needed to carry the fiction of everything-is-the-same-*but*-size forward. In 1948, Florence Moog, a zoologist, exposed this overreach with the scathing remark that "no very acute insight is needed to see that such a tremendous bulk could not be borne in a frame of human proportions," before she then concentrated on Brobdingnagian legs.[7] A gazelle's slender legs may be two-thirds of the animal's height at the shoulder; a heavy horse's only about half the height; and the round pillars supporting an elephant hardly more than a third. She concluded that: "We need no more than the zoo-keeper's rule that once around the forefront of an elephant is half the height of the body to make it clear that the delicacy of the feminine ankle must have been a matter of no interest in Brobdingnag."

Moreover, much like an elephant, such a creature would hardly have any distinct neck and could not walk on anything resembling the arched structure of human feet; most likely, it could not walk upright at all. But all of this is actually irrelevant because the creatures one-twelfth and twelve times Gulliver's height were neither that light nor that heavy. Those distortions arose from Swift's big mistake, which was to assume that the body mass (M) of organisms is proportional to the cube of their height (body length, L): $M \sim L^3$ ($\sim$ is the sign for proportionality). This would be true if all mammals were perfect and inanimate cubes or spheres, rather than complexly shaped three-dimensional organisms with intricate metabolisms who maintain steady body temperature.

Nearly 150 years after Swift published *Gulliver's Travels*, a Belgian mathematician and statistician, Adolphe Quetelet (whose work we will encounter in more detail in this book's seventh chapter), used a variety of data on weights and heights to conclude that "we will find that the weights of mature individuals with different heights are about the same as the squares of the heights."[8] The actual proportionality for human bodies is thus $M \sim L^2$: the mass scaling with the square (not with the cube) of height. Although this has been known since 1833, some recent publications dismantling Swiftian errors still cling to the impossible $M \sim L^3$ theoretical rule, concluding, for example, that the Lilliputians could not have had human-like intelligence because they would have only 1/1,728 (fewer than 0.06 percent) neurons in their brain tissue.[9] The conclusion about the impossibility of human-like sapience is correct, but (as I will show later) the calculation of brain capacity is wrong.

Quetelet's index (M / L^2) was confirmed by Ancel Keys, an American physiologist, when he conducted his famous epidemiological studies of diet (above all, fat and cholesterol intake) and mortality due to cardiovascular disease. In 1972, Keys renamed it the body mass index (BMI), and ever since the measure has been in wide use as the populations of many affluent nations have seen increasing rates of overweight and obesity.[10] To calculate this telling indicator of being overweight, simply divide weight in kilograms by the squared height in meters: the normal (non-obese, non-underweight, non-malnourished) range is 18.5–24.9. For example, a 160-centimeter tall woman weighing 55 kilograms ($55 / 1.6^2 = 21.5$) will have her BMI almost perfectly in the middle of the desirable range.

Assuming that Gulliver's BMI was similarly placed within this healthy range (21–22), his weight would have been about 67 kilograms ($22 \times 1.75^2 = 67.4$). An average Lilliputian (assuming, again, a BMI of 22) would have weighed almost exactly 500 grams, or just a smidgen less than half a kilogram. Properly scaled, an adult Lilliputian would thus have a body mass more than 10 times larger than Swift's erroneous attribution, and instead of being equivalent to a tiny shrew he would be more like an eastern gray squirrel and just a bit larger than a cotton-top tamarin (400–420 grams), a cute black-and-white Colombian arboreal primate.[11] Conversely, a Brobdingnagian adult male would weigh 9,702 kilograms (less than a tenth of the mass implied by Swift's error): at nearly 10 tons that would be the weight of an exceptionally large African male elephant but in a much taller, bipedal body.

Inevitably, the same scaling errors affect Swift's assumptions about the body mass of "six cows and two bulls alive, with as many ewes and rams" that Gulliver takes with him when he leaves Lilliput—he pulls these miniature animals out of his pocket in order to convince the captain of the ship who rescues him that the dangers he has undergone have not unhinged him mentally. If we apply the 1/1,728 scaling (with Lilliputians at 39 grams), then bulls would be at least 300 grams each, cows at least 175, and rams at about 60 grams and ewes at no less than 30 grams. The pocketed mini-herd of black cattle and sheep would weigh at least 2, more likely up to 3 kilograms: that would require some pretty capacious pockets, to say nothing about the state of jumbled animal bodies inside them. But when we fix Swift's scaling error and make the animals proportionate to the properly scaled body weight of adult Lilliputians (about 500 grams), then we have at least 4-kilogram bulls, at least 2.2-kilogram cows, nearly 1-kilogram rams, and about 500-gram ewes, and Gulliver's transported mini-herd adds up to more than 25 kilograms, obviously beyond any pocket capacity!

Leaving the livestock aside, there is no doubt that the properly scaled Lilliputian body mass of about half a kilogram is far more realistic than the mass calculated by cubing the height—but it still raises some obvious questions. Does Quetelet's rule, derived solely from adult human measurements, apply to all primates—whose extreme body weights range from 100 grams for tiny bush babies to 180–200 kilograms for the largest male gorillas? Gulliver would be 134-fold heavier (not 1,728 times) than a Lilliputian, and this multiple is easily within the just-indicated extremes of primate weight, and hence that range of scaling seems legitimate. In contrast, there has never been a terrestrial bipedal mammal whose height and mass would come close to those of a Brobdingnagian, who would be roughly 145 times (not 1,728 times) heavier than Gulliver and whose height (at 21 meters) would be at least seven times greater than that of the tallest primates.[12]

In the 21st century, the body masses implicit in Swift's original (L^3) assumption should immediately suggest some second thoughts about the behavior of such creatures to anybody interested in dinosaurs. After all, as implied by Swift, a Brobdingnagian adult male would have weighed as much as the largest-known quadrupedal sauropods belonging to the *Titanosaurus* clade. Instead, Swift assumed that both the tiny and the giant populations behaved exactly as humans—walking, jumping, riding horses, drilling armies, sailing ships. This was Swift's third—and second highly consequential—mistake, because such drastic changes in size could not happen without profound changes in the anatomy, metabolism, and behavior of those extreme organisms.

Charging Swift with this lapse could easily be seen as both unfair and misplaced: after all, his brief Cambridge education was in theology, making him, in his deprecatory self-assessment, "not the gravest of divines."[13] He was neither a mathematician (statistics had yet to emerge as a specific branch of knowledge) nor a biologist. And his book came out nine years before Linnaeus's *Systema Naturae* reviewed the diversity of species, and more than a century before Charles Darwin's book about their origins. Swift's intent was merciless satire and to expose human folly—not lessons in the as-yet-unborn science of scaling.

But, remarkably, Swift's scaling error could have been avoided if he had come across a chatty but weighty tome published 88 years before *Gulliver's Travels*: Galileo Galilei's *Dialogues Concerning Two New Sciences*.[14] Right at the beginning of the first day's dialogue, Sagredo (a layman named after Galilei's friend) says that: "since mechanics has its foundation in geometry, where mere size cuts no figure . . . [if] a large machine be constructed in such a way that its parts bear to one another the same ratio as in a smaller one, and if the smaller is sufficiently strong for the purpose for which it was designed, I do not see why the larger also should not be able to withstand any severe and destructive tests to which it may be subjected."

To which Salviati (an academician, the second of the three participants in the dialogues and Galilei's alter ego) promptly replies that "the common opinion is here absolutely wrong," and that Sagredo must discard the notion that machines built of the same material and maintaining the same ratio between parts are able to resist equally (or rather proportionally) any external disturbances: "For we can demonstrate by geometry that the large machine is not proportionately stronger than the small. Finally, we may say that, for every machine and structure, whether artificial or natural, there is set a necessary limit beyond which neither art nor nature can pass."

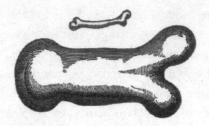

Salviati's convincing bone comparison.

Limits that apply to machines also apply to all organisms, and Salviati proceeds to sketch a bone whose length is three times the human size. The bigger bone is grotesquely massive, and hence:

> if one wishes to maintain in a great giant the same proportion of limb as that found in an ordinary man, he must either find a harder and stronger material for making the bones, or he must admit a diminution of strength in comparison with men of medium stature; for if his height be increased inordinately he will fall and be crushed under his own weight. Whereas, if the size of a body be diminished, the strength of that body is not diminished in the same proportion; indeed the smaller the body the greater its relative strength. Thus a small dog could probably carry on his back two or three dogs of his own size; but I believe that a horse could not carry even one of his own size.

Could there be a clearer and more convincing explanation of Swift's fundamental proportionality error? Let us now consider the key implications of miniaturization and gigantism applied to human-like organisms. Because surfaces scale as L^2 and volumes as L^3, the ratio of body area to body volume rises as the organisms get smaller, and tiny creatures, with relatively larger surfaces, lose their body heat much faster than the larger ones. Because heat loss is proportional to the surface area, there are no warm-blooded animals smaller than the Etruscan shrew or a bumblebee bat—both weighing just 2 grams and with bodies no more than 3–5 centimeters long. Lilliputians, more than 200 times heavier but only 3–5 times longer than these tiniest mammals, would not lose heat as rapidly but, much like all very small animals, they would lose it much faster than humans, and they would have to eat frequently—certainly many more times than three meals a day scheduled at fixed times.

Another inescapable consequence of being so small is the burden of getting wet. Because of a larger area/volume ratio, proportionately more water will cling to a small body after getting out of water or after being soaked by heavy rain. For an adult man the total would be equal to about 1 percent (an inconsequential addition); for a Lilliputian it would amount to 10–15 percent of his body weight, akin to dressing in a heavy winter coat. Or as Philip Morrison, a physicist, put it in his MIT Christmas lecture in 1968: "Lilliputians must be a hungry lot, restless, active, graceful, but easily waterlogged. You can recognize these qualities in many small mammals, like a mouse."[15]

And the Brobdingnagians? At nearly 10 tons (their unexaggerated body mass derived by Quetelet's formula), they would face the opposite thermal challenge: interior overheating of their massive bodies that would only increase with strenuous activity. As for their locomotion, the key limit is set by the material that makes up all mammalian bones: crystalline calcium apatite, the mineral that is embedded in the matrix of collagen (the bone-forming protein of all vertebrates). But there is no need to make any theoretical assumptions about maximum loads or to apply scaling exponents derived from analyses of much less massive mammals. Just imagine humans enlarged to the mass of the heaviest of male African elephants,[16] but more than four times taller, walking on two legs, and not only running but jumping into saddles and riding horses.

Gulliver says that the Brobdingnag had 32,000 horses in the army, and that "a cavalier, mounted on a large steed, might be about ninety feet high" (about 27 meters). Medium-size horses used by European armies during Gulliver's time weighed around 500 kilograms, which means that a proportionately enlarged Brobdingnagian horse would weigh 70–75 tons, as heavy as *Argentinosaurus*—so far the largest identified sauropod, whose height of 21 meters would match the height of the Brobdingnagian steed.[17] It is difficult to imagine *Argentinosaurus*-like horses that are proportionately enlarged incarnations of nimble-footed animals that can rear up on their hind legs, carry a rider, wheel around, and gallop, to say nothing about being trained to dance to classical music much like Vienna's Lipizzaners.

J. B. S. Haldane, one of the founders of population genetics, used the heights of Giant Pope and Giant Pagan (at about 18 meters, just a bit shorter than a Brobdingnagian) from *Pilgrim's Progress* to make the same scaling point. Their bones would have to support 10 times the weight borne by a unit area of human bone—but as the human thighbone breaks when loaded with about 10 times human weight, those giants, much as the unimaginable dinosaur-like horses, "would have broken their thighs every time they took a step."[18] The only way to attain, support, and move such body mass on Earth would be to become aquatic: adult blue whales, the planet's largest surviving mammals, weigh more than 100, and as much as 200, tons.[19]

Another notable size-dependent difference is the effect of gravity. As size declines so does gravity's force, while air resistance and viscosity gain in importance. Insects crawling upside down on leaves or sitting on a ceiling can ignore gravity, but they must beware of getting wet. And the fate of falling and tumbling creatures depends on their size. Resistance is proportional to the surface of the moving object, and for a small creature with a relatively larger area/weight ratio, resistance is much higher than for larger bodies. A falling insect is in absolutely no danger of being squashed on impact as it tumbles from a bush; for squirrel-like Lilliputians, much as for any small creature, a fall from a great height might make little difference; for humans, falling even from relatively small heights presents obvious bone-breaking dangers; while for large animals falls from the same height mean certain death. Or, as Haldane put it: "You can drop a mouse down a thousand-yard mine shaft; and, on arriving at the bottom it gets a slight shock and walks away, provided that the ground is fairly soft. A rat is killed, a man is broken, a horse splashes." Needless to say, a Brobdingnagian horse falling into a deep shaft would be compacted beyond recognition.

Haldane also expressed yet another fundamental point about size and complexity. A microscopic worm easily absorbs all the oxygen it needs through its smooth skin, and it has a primitive short and straight gut; large animals require complicated internal organs (human lungs have an internal surface area of 30–50 square meters). Hence Haldane's memorably correct contrast: "The higher animals are not larger than the lower because they are more complicated. They are more compli-cated because they are larger. Just the same is true of plants."

The simplest plants—the smallest algae—are just tiny round cells in water or attached to various surfaces; trees—the largest plants—have to support their extensive radiation-capturing surfaces (otherwise known as leaves) with sturdy trunks. That is why Swift's basic everything-is-the-same-*but*-size assumption would not work any bet-ter for crops in his undersize and oversize worlds than it would for animals—and I will use wheat, the staple of Gulliver's society, as per-haps the best example of the consequences of changing size.

Mature, modern, short-stalked varieties of spring wheat are typi-cally around 70 centimeters tall, the result of generations of breeding aimed at redistributing the plant's above-ground biomass.[20] In modern cultivars, about half of it is in harvested grain, half in stalks and leaves. In contrast, traditional varieties of this annual herbaceous plant were tall, often even exceeding 1.5 meters and yielding up to four times as much straw than grain in weight.[21] Assuming that in Gulliver's time English wheat was 1.3 meters tall, in Lilliput the proportionately sized plants would be no more than 11 centimeters tall and in Brobdingnag they would rise to more than 15 meters, the height of a five-story house (for comparison, average tree height in the US is 23 meters).

No annual herbaceous staple grain plant, whose stems dry out after one season of growth, can be that tall. The tallest grain crop cultivated on a large scale, American corn, averages 2.4 meters by midsummer.[22] As plants grow taller, less of their dry matter is allo-cated to expanding leaves and more of it goes into structural tissues, above all into stems. This is the price paid for larger size: as individual plants compete for access to solar radiation, soil nutrients, and water, each enlargement of the leaf area must be accompanied by a propor-tionate increase in the mass of structural tissues required to support new photosynthetic surfaces. Wheat plants 15 meters tall could not have straw stems but would have to be more like bamboo, a tall per-ennial grass with thick hollow stems.

Moreover, it would be impossible to have anything like wheat's plant density on a Brobdingnagian scale. Standard practice is to have about 3.5 million viable wheat stems per hectare (350 per square meter, or one plant for roughly every 5-centimeter square)[23]—but as individual plants get taller, their continued growth can be accommodated only by reducing the number of stems per unit area, and "self-thinning" (plants dying off to make room for more successful neighboring survivors) regulates the further growth of tall stems. Modern varieties of ground-hugging grasses (less than 20 centimeters tall) have more than 10 million stems per hectare; 70-centimeter tall wheat (as just noted) has about 3.5 million plants per hectare; 2.5-meter-tall corn field averages just 75,000 plants—but there would be only some 400 bamboo clumps (or bamboo-like giant wheat) per hectare.[24]

Calculations done to expose the fallacies of everything-is-the-same-*but*-size can go on and on. What kind of force would be needed to pump blood to the head of a Brobdingnagian taller-than-a-dinosaur horse—or to cut 15-meter-tall mega-wheat whose bamboo-like stems would be at least 15 centimeters across? Could any heart actually do such a job? Would a proportionately sized iron sickle work? Obviously, Swift's imagined journey has furnished us with an opportunity to question, and to set straight, many constructs that are fine for riveting fiction but untenable in reality, and we can use Gulliver's extreme experiences as instructive examples of the fact that size cannot change alone. Everything else has to change too.

And so we should be doubly thankful to the dean: for what he got right, and for what he got wrong. For providing us with a perennially fascinating and entertaining combination of adventure, caricature, political allegory, social critique, merciless analysis, and probing commentary on the human condition—as well as with, unwittingly, such rich and apposite material for illustrating the errors that arise from neglecting the effects and consequences of scaling in organisms.

And, still, I am not done with the Dublin dean. I have to tackle yet another fundamental miscalculation, by taking a closer look at what is perhaps the most consequential of all size-dependent scaling relationships: the change of metabolism with changing body mass. But the road to rectifying that mistake began with the rise of a new line of scientific inquiry.

A brief history of allometry: of skin and crab claws

How do some essential structural (bones) and functional (skin, internal organs) properties of organisms scale with size? When did we begin to investigate these relationships in systematic ways? As in many other cases of new scientific inquiries, the first quantitative attempts at scaling human organs began in Germany in the second half of the 19th century. In 1879, a German physician, Karl Meeh, published a lengthy report on quantifying the size of body surface as the function of total body mass.[25] Unlike with internal organs, whose weights or volumes are fairly easy to measure during autopsies, finding the total area of skin was an immense challenge during the 19th century.

Meeh proceeded by marking shapes on bodies, copying those shapes onto tracing paper (and calculating the areas, either geometrically or by weighing the paper), and wrapping the cylindrical areas (legs, arms, torso) in strips of millimeter-wide paper. He examined only 16 subjects (10 children and 6 adults), ranging between 6 days and 66 years of age—clearly too small a sample to derive any solid conclusions. But the formula he developed to generalize the dependence of surface area on a body's weight (the latter growing as the former raised to the power of 2/3) is still in use nearly 150 years later. By 1915, two American physicians, the DuBois brothers, introduced their formula that used both weight and height to predict body surface area; remarkably, this quantification has remained in use ever since, even as more than 20 other equations were published between 1916 and 2010.[26]

In 1892, Otto Snell, another German physician, was the first researcher to address the scaling of a body organ whose mass increases disproportionately with increasing body mass.[27] Snell's organ of choice, dictated by data availability, was the human brain—whose size increases at a slower rate than the body mass of an entire organism. This means that the brain's share of total body mass becomes proportionately smaller as the body weight of mammalian species increases. Snell presented data for 20 mammals and 25 birds but offered no predictive generalizations. His mammalian series confirmed the expected decline in brain-to-body ratio, with the extremes ranging from 1/23 for a common shrew to 1/22,000 for a bowhead whale.

Five years later, in 1897, the Dutch naturalist Eugène Dubois published a lengthy report about brain size as it relates to the body size of mammals. He expressed the relationship in a simple general formula, $e = cs^r$: e for encephalon, the weight of the brain; c for the coefficient of cephalization; s for the body mass; and r for the coefficient of relation.[28] This was the first time a scientist had used what is known as a power-law function to calculate an actual mass-dependent variable: it computes how a change in one quantity results in a proportional change in another. Enlarging a square is the simplest example of a power-law: doubling the length of its side, x, will quadruple its area, y, or $y = x^2$.

Just a year later, a young French physiologist, Louis Lapicque (1866–1952), began to use Dubois' formula for the comparison of relative weights of brains—both within the same species (starting with dogs) and between different species. In 1907 Lapicque published a graph plotting brain weights in related animals (including rodents, grazers, and primates) and their body masses: it clearly showed that larger animals do not have proportionately heavier brains.[29] Dubois and Lapicque thus provided the lasting foundations of modern scaling studies: a power-law formula to calculate the value of size-dependent variables, and a revealing, graphic way to illustrate the scaling progressions.

As is often the case in the history of scientific advances, Dubois and Lapicque, both writing in French, do not feature in the standard English-language narratives of scaling history. Those are dominated by the work of Julian Huxley, an honorary lecturer in experimental zoology at King's College London and a grandson of Thomas Henry Huxley, the leading public advocate of Darwinian thought. Huxley's first scaling publication was in 1924, when his short paper—on what he called the "heterogenic development"—addressed the growth of the extraordinarily large right claws of male fiddler crabs.[30] Huxley measured the lengths of crabs' growing bodies and the lengths of their outsize right claws, and when he plotted the logarithmic values of the two variables he got a remarkably straight line whose slope clearly indicated that the claws were growing faster than the rest of the body: its angle of ascent was steeper than the angle of the body-growth line.

Large male fiddler crab claw.

In 1932, Huxley published his lengthy inquiry into the problems of relative growth, because he considered the previous efforts sparse and rather disconnected.[31] His name became strongly associated with power-law scaling, mainly because four years later he and Georges Teissier gave the enduring name to this new field of study: in a short paper in *Nature* they chose the term "allometry" (*allos* = other, *metron* = measure; the measurement of otherness).[32]

Expressed in universal terms, the scaling (power-law) equation is $y = cx^r$, where y stands for the size of the studied (dependent) variable, x for the independent variable, r for the exponent that governs the relationship, and c for a constant multiplier required to obtain the result in specific units (of mass, length, area, etc.). The case of male fiddler crab claws, growing much faster than the rest of the body, was one of a hyperallometric relationship (*hyper* = over, above)—as opposed to a hypoallometric (*hypo* = under, below) sequence where an organ would be growing at a slower rate than the total body mass.[33]

But none of Huxley's examples of allometric scaling became as famous as the relationship between metabolism and animal weight that was quantified by Max Kleiber—a Swiss biologist who worked at the California Agricultural Experiment Station—in a report published in 1932, the same year as Huxley's book.[34]

Metabolic scaling is the subject of the next chapter, and in the remainder of this one I will look at other structural and functional variables related to weight (body mass). Why mass? Any size variable would do, but such linear measures as height, length, width, or circumference require specific definitions. For example, the height of a horse is not measured from the ground to the top of its head but to the highest point of its withers—the ridge between the animal's shoulder blades—while for many animals, length measurements must be specified with or without tails, and in most cases they are harder to ascertain than body mass.

That is why allometric studies have always related many physiological (and later also various ecological) variables to a less tricky variable: body mass. The basic all-purpose scaling equation can be written simply as $y = cM^r$, where y is the variable of interest (size of an internal organ, size of a home territory), c is, again, a constant multiplier (used to express the outcome in specific units), M is the body mass of an organism, and r is a scaling exponent. When $r = 1$ we get a simple linear relationship, as a studied variable scales isometrically (*isos* = equal)—that is, the mass (or the volume) of an organ changes at the same rate as the changes in body mass—and plotting these variables on linear axes gets us a straight line.

If a 1-kilogram animal has a 10-gram (1 percent of the total) internal organ, it would still be 1 percent of the total in a 10-kilogram animal (10 times heavier) but it would weigh 10 times as much (100 grams), etc. This isometric scaling is not common: among the essential organs only lung weight, vital capacity (the greatest volume of breath that can be exhaled after the deepest possible air intake), and blood volume—and, according to some studies, hearts and the total mass of body fat—scale isometrically or come very close to that progression. As already noted, hypoallometric scaling means that organs or functions change at a slower rate than body mass, and the opposite is true (organs change at a slightly faster rate than body mass) in the relatively uncommon cases of hyperallometric scaling.

The brain has remained a favorite object of scaling investigations,

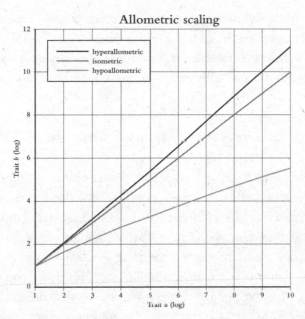

Isometric (1:1) and hyperallometric scaling (>1) of organs is uncommon. Hypoallometric scaling (<1) is encountered most frequently.

and studies show exponents as low as 0.70 and as high as 0.81, with the most comprehensive examination of about 1,500 mammalian species showing that the exponent for their brain growth is 0.75 ($M^{0.75}$). As body weight doubles, for example, from 10 to 20 kilograms (that would be the difference between a colobus monkey and a Canadian beaver), brain weight will increase only 1.68 times. As first graphed by Lapicque, allometric scaling results in captivating, and informative, illustrations. When you plot body weight on the horizontal axis and an organ weight on the vertical axis, a hypoallometric scaling will follow a curve bending gently downward, clearly indicating that the weight of an organ is increasing at a slower rate than body mass. But when you express all of these values in logarithms and plot them on two logarithmic axes, you get—as in the pictured example—a straight line.

In reality, some specific values of actual sizes of an organ or a function will fall right on, or very close to, the line calculated by a particular allometric equation; other data points will be noticeably below or above it. In the nearly 150 years that have elapsed since Meeh's pioneering skin-area investigations, the science of organismic scaling has broadened and deepened but—as the next section will show—it is still trying to fit complex empirical evidence into generalizing formulas, and it is still dealing with exceptions and uncertainties.

Scaling of organs: brains, hearts, bones

That larger bodies will have larger organs is obvious, but does Lapicque's discovery of brain mass increasing at a slower rate than the total body mass hold for all organs—or are they growing at the same rate as bodies, or perhaps even faster? Calculations of body-size-dependent skin area are one of the best examples of these lasting uncertainties: comparisons of available formulas show considerable ambiguity, with exponents ranging from 0.38 to 0.66. As a result, most of these body surface area formulas are in good agreement only for a very limited range of body weights, but even then it is impossible to say which formula should be seen as the most accurate.[35]

This matters because body surface area should be known with as much accuracy as possible in order to provide some cancer chemotherapies, transplants, and burn and poisoning treatments. Scaling formulas show a fairly high degree of agreement for the most common adult weights (50–100 kilograms) and heights (150–170 centimeters), but result in unacceptably large differences (as much 20–30 percent) for children and for heavier adults. The largest discrepancies are associated with low heights, and the absolute differences are high (as much as half a square meter) even for very tall (2-meter) individuals. Applied to Gulliver's worlds, the traveler's body surface area (using a recent equation combining weight and height) would be about 1.8 square meters, and the same formula would result in about 86 square meters of skin for a Brobdingnagian adult male: he would weigh nearly 145 times as much as Gulliver but his body surface area would only be about 48 times larger.

But ambiguity of available formulas for calculating body surface area is not even the most fundamental uncertainty: that comes from the assumption, implicit in all of body surface area studies, that the skin is flat, as if pores and follicles were nonexistent and do not contribute to the total area. But, of course, they do, their density per unit area is different in adults and children as well as among different populations, and these structures are metabolically highly active, with pores being the dominant enablers of evaporative cooling. Consequently, all extant mass-dependent—or mass- and height-dependent—formulas are incomplete.

Turning to brains, the linear fit can be used to see which animals are more or less brainy than expected. Among the mammalian orders, primates, carnivores, some tree shrews, and toothed whales have larger brains than expected, with primates (especially humans and chimpanzees) being significantly above the mammalian line.[36] In contrast, the brains of dogs, horses, and elephants are as large as the expected (calculated) values, while the brains of hedgehogs, pigs, hippos, lions, and also blue whales (the largest mammals) have smaller-than-expected brains.

The size of the human brain has been the topic of hundreds of books and tens of thousands of papers in scientific journals.[37] Brains that are extraordinarily larger than expected for a mammal whose

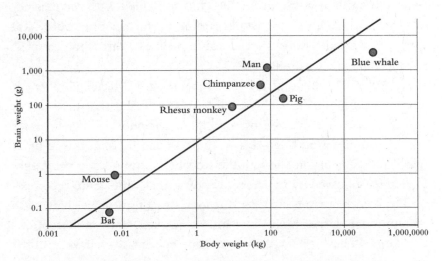

Primates have significantly heavier brains than the mean expected weights for vertebrates; the same is true for dolphins and for some species of mice, but bats, pigs, and blue whales are relative laggers.

body weight range overlaps with chimpanzees and pigs are, of course, the pinnacles of human evolution (alongside our bipedalism, dexterous hands, and a remarkable combination of relatively low demands for water and high sweating rates that make it possible to acclimatize to hot environments). Adult pig brains weigh less than 200 grams; adult chimpanzee brains no more than 450 grams; while the average for adult men (taken from a study of 18- to 35-year-old males who died of sudden traumatic deaths) is 1,407 grams, with the range of 1,179–1,621 grams including 95 percent of all measured cases.[38]

But nothing is ever simple when dealing with such complex organs as brains over such long evolution spans. There is no doubt that primates in general and humans in particular have larger brains and hence a much larger number of nerve cells than other mammals with similar brain size, and that mammalian intelligence correlates best with the numbers of neurons and synapses and with information-processing speed. The human brain has more than 80 billion neurons (chimpanzees have less than 10 percent of that number) and about 100 trillion synaptic connections, numbers that have counterparts only in the counts of stars in assorted galaxies.[39] On the other hand, the approximate tripling of the hominid brain since the end of the Pliocene (2.58 million years ago), which has coincided with the mastery of fire and the use of more sophisticated artifacts (stone tools; later spears, bows and arrows) should not be seen as a direct indicator of enhanced mental capabilities.

As Australian anatomist and anthropologist Maciej Henneberg notes, equid brains have also tripled in size during that period, but there has been no increase of intelligent behavior among horses.[40] Increases in hominid brain size were paralleled by increases in body size—and (starting in the late Pleistocene, about 30,000 years ago) they were followed by 10 percent decrease that was, again, accompanied by a decrease in body size. But something must have changed to accommodate larger brains in same-size bodies, and the exceptionally large brain size/body ratio of our species may have resulted primarily from the reduction of the gastrointestinal tract and its musculoskeletal support. This trade-off was made possible by better, more meaty diets and more palatable processed (cooked, milled) food that could be metabolized by a smaller digestive tract. Our mental abilities may thus owe more to a functional rather than an anatomical evolution.

And brains in Gulliver's world? Even if we assume that the Lilliputian brains would be proportionately as large as human brains, they would weigh about 10 grams: such a tiny brain would have about 800 million neurons compared to the 86 billion neurons in the human brain, implying merely 0.9 percent of our cognitive and learning capacity.[41] Would such brains be capable of forming and running a society whose intricacies and intrigues duplicate human-like complexity? The answer is obvious. In contrast, assuming that adult Brobdingnagian brains would be proportionally as massive as are human brains, they would weigh close to 200 kilograms—and the obvious question would not be about their superior reasoning capabilities but about the weight of the skull required to contain such a massive organ, and about the overall upper-body skeletal structure, diameter of the neck, and the strength of the muscles necessary to keep that mass upright and able to rotate it.

Turning to other major organs, some studies have shown that heart sizes scale isometrically with bodies; others indicate that such scaling does not apply across the entire mammalian range, from shrews to elephants.[42] For example, the adult dog heart is a larger share of a husky's total body weight (about 0.7 percent) than a heart is in humans (0.4 percent), while in South African antelopes, relative heart mass decreases from 6 grams per kilogram for a tiny 12-kilogram duiker to less than 4 grams per kilogram for the large (500-kilogram) eland.[43]

Cardiac output in mammals is a more complicated matter. In those weighing up to 10 kilograms it is isometric; for medium-sized mammals, including humans (10–100 kilograms), the exponent drops to 0.71; and for larger mammals it further decreases to 0.67.[44] Increasing mammal size brings a reduction in cardiac output per unit of body mass, while systolic blood pressure remains relatively invariant with size (120 mm Hg for both a mouse and a horse).

The human heart (0.25–0.35 kilograms) has an aortic diameter of 3 centimeters. The largest mammalian heart (450 kilograms, belonging to the blue whale) weighs 1,300–1,800 times as much, but its aortic diameter is only about eight times larger (23 centimeters); this means that in large bodies, the heart is throttled by a relatively small aorta.[45] Obviously this puts an important limit on mammalian body size. No less obviously, what is possible in a 100-ton horizontally oriented animal buoyed by water would be impossible in a similarly sized terrestrial biped: hearts in Brobdingnagian bodies implied by Swift's original calculations are beyond any functional scaling.

All animals have approximately the same number of lifetime heartbeats, which means that the hearts of smaller and short-lived creatures must beat much faster than those of massive mammals (the mass-dependent scaling exponent is −0.25), with frequencies (beats per minute) ranging from more than 1,000 for tiny shrews to less than 30 for the largest male elephants—and less than 10 for blue whales.[46] Lilliputian hearts would beat more than 200 times per minute, while adult male Brobdingnagian hearts (in bodies whose weight is properly adjusted to be less than 10 tons) would beat less than 20 times a minute. The human average (72 heartbeats a minute, with the most common range being 60–80) overlaps with the lower range for adult pigs, a similarly massive mammal whose heart is only slightly larger than an average human heart, making the organ the best substitute for training cardiac surgeons.[47]

And what about the question posed at the beginning of the chapter regarding the suitability of heart transplants? Suboptimal sizing of transplanted hearts is clearly undesirable. The standard practice is to limit the pool of possible donors only to those whose body weight differs by less than 30 percent from the recipient's weight, but because adult male hearts are, on average, about 35 percent heavier than female hearts, the largest study of transplant histories showed that men receiving female hearts were 32 percent more likely to die in the first year after receiving a new heart, and the cause is entirely due to suboptimal sizing that forces smaller hearts to strain excessively.[48]

Livers, kidneys, and digestive tracts are among the much-studied internal organs that usually scale hypoallometrically (changing at a lower rate than body size), but in some cases isometrical scaling (changing at the same rate as body size) cannot be excluded. In turn, the hypoallometric scaling of energy-expensive internal organs makes a decisive contribution to the hypoallometric scaling of whole-body metabolism. The heart, kidneys, liver, and brain make up only 5.5 percent of adult human body mass (0.5, 0.4, 2.6, and 2 percent respectively) but are responsible for nearly 60 percent of the resting metabolic rate (respective rates are 8.7, 8.2, 21.6, and 20.2 percent).[49] In contrast, skeletal muscle makes up about 40 percent of adult (non-obese) male body mass but consumes only 22 percent of the energy at rest, with fat tissues (21 percent of body mass) using the rest.

The exponent for skin mass is (depending on the mammalian data sets used for analyses) as low as 0.84, but the scaling can be hyperallometric, with the exponent larger than 1.0. That is also the case for mammalian skeletons. Heavier skeletons are expected, as we have seen, because in order to have the same structural strength larger animals must have bigger bones, with different studies suggesting scaling exponents between 1.09 and 1.14.[50] The latter rate would mean that a 100-kilogram animal would have a skeleton not twice but at least 2.15 times heavier than a 50-kilogram mammal, and a hippopotamus 10 times as heavy as a domestic pig would have a skeleton more than 12 times as heavy. Very similar scaling is true for the total weight of body fat, with reported exponent values between 1.14 and 1.19. To sum up, smaller animals carry relatively more skin and have relatively larger brains, hearts, kidneys, and livers, while larger animals have heavier bones and are more fatty.

Mammals have a substantial variation in muscle mass: depending on the mammalian order, the shares as a percentage of total body mass range from just 22 to 61, with primates (22–49 percent) at the lower range of this spectrum. But despite these substantial differences, their muscle mass and body mass scale isometrically, and primates have a slightly higher scaling coefficient than non-primate mammals (1.05 and 0.99).[51] Gender differences in muscle mass are obvious: a study of 18- to 88-year-old adults showed skeletal mass among men to be significantly larger than in women both in absolute and relative terms (33 vs. 21 kilograms; about 38 vs. 31 percent); the differences were greater in the upper than the lower body, and a noticeable decrease in absolute skeletal mass starts near the end of the fifth decade.[52]

The last major organ whose hypoallometric scaling I will review is the eye.[53] There is an added benefit of not only having a large eye but also an anatomically long one. A greater distance between the cornea (the eye's protective outer layer) and the retina (the innermost, light-sensitive layer) increases the size of the received image, and that is advantageous both for finding food and eluding predators. Birds of prey have proverbially superior eyes: eagles have 20/4 vision (seeing from 20 feet what a person can see from 4 feet), and they also see colors more vividly and can even sense ultraviolet light.

Analysis for all vertebrates and independent analyses for all fishes, reptiles, birds, and mammals show that birds and primates have relatively large eyes, rodents and reptiles have relatively small eyes, and that the variability of fish eyes does not lend itself to any general conclusion. Brobdingnagians would have no problem with having large-size eyes set in their enormous heads (as relative eye size diminishes with body size), but in order to have human-like vision Lilliputians would have to have disproportionately large eyes that would, in relation to their face, look more like those of bush babies than similarly sized small squirrels or guinea pigs.

From the 1950s, as the new science of animal ecology made enormous advances based on painstaking observations of animal behavior in the wild, biologists began to quantify the effect of body size on key parameters of existence. None is more important than the size of home range: the area of land that contains all the resources required for an animal's survival and reproduction.[54] A comprehensive study of the home ranges of more than 500 vertebrate species resulted in some expected findings. Birds of prey have the largest home ranges; carnivorous mammals have to roam more than herbivores; fishes have the smallest ranges for their size. The ranges of carnivorous mammals scale hyperallometrically with their mass, and more steeply than those of herbivorous mammals: carnivores weighing 1 kilogram (the mass of a European polecat or a small fennec fox) have home ranges about 14 times larger than herbivores of the same size.[55]

Similarly, the home ranges of carnivorous birds (from small falcons

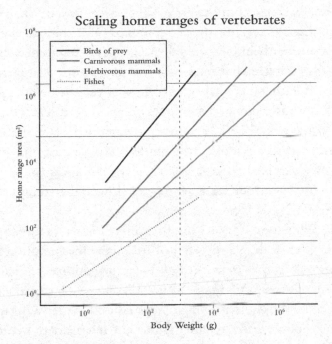

Scaling home ranges of vertebrates

Birds of prey have by far the largest ranges for their body mass; range sizes are intermediate for terrestrial carnivores and herbivores, and smallest for fishes.

to large condors) scale higher and more steeply than those of herbivorous species. Also not surprisingly, locomotion is a major determinant of home-range size, and it explains more variation among species than mass alone. For a given body mass, home-range sizes increase in tandem with the feeding strategy (from ambushing to interrupted searching to near-constant cruising). Everything else being equal, the fastest animals have the largest home ranges, and flyers and runners are predicted to have home ranges, respectively, 100 times larger and 10 times larger than swimmers.

Terrestrial mammals exploit their ranges mostly by running, and analyses show a clear dichotomy between the running of small (<1 kilogram) and larger (>1 kilogram) animals. There is no detectable difference in body-mass-dependent scaling efficiency within each of these groups, but small animals are generally less efficient runners than the larger ones.[56] Why the small ones are so inefficient is not entirely clear, but the efficiency of large ones clearly owes to the reuse of briefly stored elastic energy. When a leg (be it of a kangaroo, impala, or marathoner) hits the ground, most of its kinetic energy is not lost permanently but is stored fleetingly as elastic strain energy in tendons (they can stretch by as much as 10 percent without breaking) and returned with high (>90 percent) efficiency in their elastic recoil. Just think of those Serengeti antelopes you have seen bouncing high after every recoil as they flee from a pursuing lion or cheetah. This useful capability is obviously size-limited: you cannot imagine the tendons of an elephant-like but bipedal Brobdingnagian recoiling like those of a gemsbok or a white-tailed deer that jumps 6-foot (1.8-meter) fences with ease.

At this point some readers might think that I have already said too much about scaling and size. I called on fascinating unintended scaling lessons from a renowned 18th-century author of fiction; I introduced the work of many famous scientists; I explained the results of numerous human and animal studies; I looked at how sizes of different body organs, ranging from skin and heart to brains and eyes, have changed with evolution and how they relate to the changes of body mass; and I provided basic equations so that readers can make their own calculations. Perhaps the single most important conclusion conveyed by these inquiries is how all organisms operate within pre-dictable mass-dependent limits.

But I am still not done with scaling, and next I will deal with what is arguably the most important allometric story: how does metabo-lism (the energy consumption required for survival, reproduction, growth, and activity) change with changing body size? Moreover, this metabolic scaling has applications beyond organisms: many machines are subject to it within equally narrow boundaries. Much as carbohydrates, lipids, and proteins energize organisms, fossil fuels and electricity energize combustion engines and electric motors. Does the "metabolism" of these inanimate energy converters scale similarly to the metabolism of mammals, or do those artifacts scale differently? And what about the relationship between energy use and the size of an economy? All of that is for the next chapter.

VI   METABOLIC SCALING

The growth, reproduction, and functioning of all organisms are enabled and limited by their energy intake, and so the scaling of metabolism is the most fundamental mass-related dependence to explore. Energy intake can be measured in absolute terms (joules or calories per unit of time) or related to body mass (joules or calories per gram or kilogram of body weight). Investigations of metabolism can focus on a number of variables, ranging from baselines—basal (or resting) metabolic rates determined by measurements under carefully defined conditions—through typical daily energy needs (usually called field metabolic rates in studies of wild animals), to maxima that set the upper range of sustained metabolic rates for those doing physically demanding jobs, such as lumberjacks or fishermen, or for peak short-term exertion in sports or freestyle climbing.

Gulliver's fictional worlds again provide a useful illustration. I have already explained how Swift's ignorance of Galileo's *Dialogues Concerning Two New Sciences* allowed him to describe the impossible: tiny Lilliputians and towering Brobdingnagians behaving exactly as humans. Here I will charge the Dublin divine with a no-less-mistaken scaling assumption committed by improperly relating the mass of bodies to their metabolic needs (food intake). But I hasten to add that, unlike Swift's first two scaling lapses exposed in the preceding chapter, this mistake is much easier to excuse. In the early 1720s, when Swift began writing his unusual travel book, science was still more than a century away from systematic examinations of metabolic scaling. Galilei's 1638 dialogue on scaling, almost certainly unknown to Swift, was concerned with structural integrity (animal and human bones, materials) not with metabolism, and hence in the early 1720s there were no published accounts that could be readily consulted in Dublin or that could be sent to Ireland by one of Swift's English friends with whom he corresponded.[1]

Recall that when assuming Lemuel Gulliver's height of 175 centimeters, an average Lilliputian would be about 15 centimeters tall and an average Brobdingnagian 21 meters tall; and estimating the traveler's weight to be about 67 kilograms, an average Lilliputian would weigh just 500 grams and a Brobdingnagian would come to 9,702 kilograms. Swift's great error was to believe that metabolism is proportional to mass, and given his greatly exaggerated mass/height assumption $(M \sim L^3)$ this meant that, as Gulliver explains, "His Majesty's mathematicians . . . concluded, from the similarity of their bodies, that mine must contain at least 1724 of theirs, and consequently would require as much food as was necessary to support that number of Lilliputians."

Alas, this rests on false foundations, but the correct proportion actually makes the Lilliputians better off. I have already noted the algebraic error: the actual multiple is 1,728 (not 1,724). But that small error is of no great consequence, because in reality Gulliver would need considerably less food than the king's mathematicians had calculated—and, likewise, Brobdingnagians would have to eat much less than Swift believed. Most people reading Swift's book would never notice this error. That is not surprising: modern societies may pay a great deal of attention to food and diets but a lifelong experience has taught me that too few people now get properly schooled even in basic science. Reader, thank you for bucking the trend and getting ready to understand 3/4 scaling!

Scaling of metabolism: what it takes to keep us alive

Mammalian metabolism (and hence, we must assume, the metabolism of those imaginary human-like shortened or extended organisms) does not scale with the cube of body mass (M^3). Nor does it scale with its square (M^2), and it is not even proportional to the total weight (M). To get through an average day, a burly Canadian lumberjack weighing 100 kilograms does not need twice as much food as his equally hardworking 50-kilogram Filipino counterpart. The actual exponent governing human and all mammalian mass-related energy requirements is considerably less than 1.0—and how we got to the actual fractional number requires a brief look at the history of metabolic scaling.

Its beginnings can be traced back to a brief statement in a paper submitted to the Académie royale de Médecine by Pierre-Frédéric Sarrus, a professor of mathematics, and Jean-François Rameaux, a physician, in 1838. They concluded that "all things being equal, bodies of the same nature lose at every instant quantities of heat that are proportional to the extent of their surface area."[2] This means that a three-dimensional system (body mass or volume) that supplies heat is proportional to the cube of its length ($M \sim L^3$), and a two-dimensional system (body surface or area) that emits heat is proportional to the square of its length ($A \sim L^2$). And so the area of animals, and hence their heat loss and energy demand, is proportional to $M^{2/3}$. Concurrently, the oxidative breathing, which leads to a body's heat production, must decline by a 2/3 power as the organisms grow larger: Gulliver, much like all adults, had to breathe 12–18 times a minute, Brobdingnagians only 5 times, and Lilliputians more than 150 times.

The first experimental confirmation of this 2/3 law only came in 1883, as Max Rubner, a German physiologist, verified the relationship by metabolic measurement in just seven dogs.[3] Despite a 10-fold difference in mass (the smallest dog weighed just 3.2 kilograms, the largest one 31.2 kilograms), their heat loss and hence their energy requirement per unit of body surface area was nearly identical. This was highly suggestive of a rule that might be valid for all warm-blooded organisms, but the small sample of a single animal species offered no statistically and systemically reliable foundation for generalizing this scaling. Still, in 1901 the notion received further confirmation from Carl von Voit, the doyen of German physiologists, whose small table for seven species showed that basal metabolic rates (BMRs) of horses, pigs, people, dogs, rabbits, and geese were all close to 1,000 kilocalories per day per square meter of their body's surface.[4]

Despite some questions about its universal applicability, Rubner's surface "law" of metabolism remained the norm until 1932, when Max Kleiber used data for basal metabolic rates of 13 warm-blooded organisms (including two steers, a cow, a woman and a man, a sheep, two dogs, two rats, a hen, a pigeon, and a ring dove) to show that their energy requirements scale not as $M^{2/3}$ but as $M^{3/4}$.[5] But getting representative values of an organism's BMR requires repeated measurements—in order to achieve fairly small individual variations—under a uniformly defined clinical setting.

Because BMRs are elevated not only by any activity but also by food digestion and by changes in ambient temperature, they must be measured after a prolonged period of rest, lying down in a post-digestive state (after prolonged fasting, at least for 10–12 but preferably 24 hours) in a temperature-controlled (thermo-neutral, no shivering, no sweating) room and in a state of mental relaxation. As you might imagine, this is done much more easily with paid human volunteers—university students have been frequent subjects of such studies—than with very small children or very old adults, and it is even more difficult to get representative readings for many kinds of small and agile or large and cumbersome-to-handle mammals.

Kleiber was aware of some of the inherent problems with comparing BMRs of different animals ("24 hours after the last food . . . is not the same for the steer as for the hen or the rat") and admitted that not all of his measurements were obtained under the same conditions. But he had enough confidence in his results to conclude "that the surface law should be replaced by a weight-power law," and he correctly stated that "a power function of the body weight gives a better-defined unit for measurement than the unit of body surface."

When you plot the metabolic rates calculated by Kleiber for 13 animals on linear axes—with body weights on the horizontal and metabolic rates (energy per 24 hours) on the vertical—you get a curve bending downward, clearly indicating that metabolism is increasing at a slower rate than body mass. With modern electronics (a calculator or Excel) it is easy to walk in the steps of allometry's pioneer. The heaviest animal studied by Kleiber—a steer—weighed 679 kilograms; just punch in this value and then hit the "log" key to get 2.83 (x-axis value); the animal needed 8,274 kilocalories per day, and the logarithm of that is 3.92 (y-axis value). The only man included in the set weighed 64.1 kilograms and needed 1,632 kilocalories, and the respective logarithms are 1.80 and 3.21. Finally, do the same for a ring

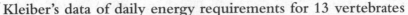

Kleiber's data of daily energy requirements for 13 vertebrates

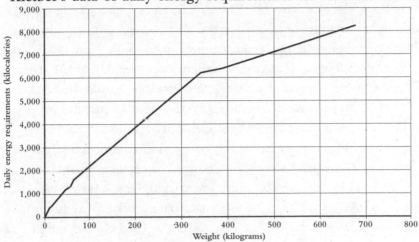

Kleiber's data of daily energy requirements for 13 vertebrates

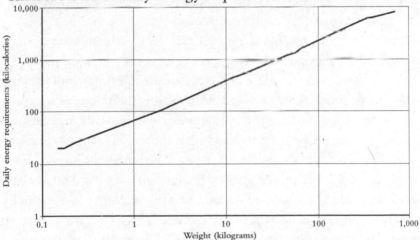

Body weights and energy requirements from
Kleiber's data table plotted on linear and logarithmic scales.

dove, Kleiber's lightest included animal (0.15 kilograms = −0.82; 19.5
kilocalories = 1.29).

Plot the three points (defined by their x- and y-values) and they
will form, almost perfectly, a straight line: you have just recreated
Kleiber's dove-to-steer line of metabolic scaling. Its slope is very

close to 3/4 (0.75): the metabolic rate of an animal increases as 0.75 power of its body mass ($M^{0.75}$)—that is, for every four log units of additional weight there will be only three log units of additional basal metabolic energy requirement. Even with just 13 original points, not all of Kleiber's measurements were exactly on the straight line: on his original hand-drawn graph you can see that the hen was a bit below and the steer clearly above the line. As more animals get added, we must expect that the values will become more scattered. If the scatter is fairly tight and if there are not too many points, you can draw a line of best fit simply by passing it through as many points as possible while leaving the same (or roughly the same) number of points above and below it. Statisticians have a better way to do that, using a procedure called the least square method to find the line of best fit for large numbers of measurements, even those that are quite obviously scattered.[6] This process also determines accurately the line's slope: actually, Kleiber's original best fit showed $M^{0.74}$ and he eventually rounded that very slightly upward to $M^{0.75}$.

As new, more reliable measurements of metabolic rates became available, Kleiber's animal range was extended from the mouse to the elephant. A 4-ton elephant weighs 200,000 times more than a 20-gram laboratory mouse, but its total metabolic rate is less than 10,000 times higher than that of a mouse, which means that per unit of weight a small mouse needs to ingest more than 20 times as much energy as does a medium-size elephant. The metabolic rates of hundreds of additional animal species, including reptiles and birds, were gradually added to Kleiber's small set, the scaling was eventually extended to organisms whose body masses range over more than 20 orders of magnitude, and the 3/4 "law" became one of the most-cited scaling relationships in life sciences.

Applications of Kleiber's "law" showed that the exponent works fairly well for the resting metabolic rate in adults. If you want to get energy requirements in kilocalories per day, then all you have to do is multiply the scaling result by 70: $Ekcal = 70M^{0.75}$. If you want to get a daily metabolic rate in watts, then the multiplier constant changes to 3.38. This means that Gulliver's basal metabolism would be about 1,640 kilocalories a day ($70 \times 67^{0.75}$) or almost 80 watts. This is a useful yardstick: an adult man's basal metabolism proceeds at a rate 20 percent lower than the energy consumed by a standard 100-watt incandescent light bulb.

In order to get actual daily energy requirements, we have to multiply BMRs by typical physical activity levels (PAL). For adult men, the multiple of 1.55 should be used for light activities, 1.78 for moderate exertions.[7] Gulliver's activities—exploring new worlds, walking around, entertaining the natives by showing off his fencing skills—would fit mostly between these two rates, that is between about 2,500 and 2,900 kilocalories a day. In contrast, the resting metabolic rate of adult males in Lilliput would only be about 40 kcal/day ($70 \times 0.5^{0.75}$), but given their necessarily more active life (at least doubling their BMR) they would need close to 80 kcal/day.

Consequently, on his low-activity days Gulliver would require a daily food supply equivalent to that of only about 30 Lilliputians. On very active days—such as one spent hooking up and pulling the entire invasion-poised Blefuscudian fleet to Lilliput while wading through the sea—the equivalent could be as much as that of 40 Lilliputians. In either case it is only a small fraction of the total that resulted from Swift's erroneous scaling of daily energy needs determined by cubing the body mass. Given this reality, there is no danger that Gulliver's accidental sojourn would overstrain Lilliput's agricultural output. After all, Mildendo, Lilliput's capital, is "capable of holding five hundred thousand souls," so feeding the equivalent of an additional 30–40 people (the maximum addition amounting to less than 0.01 percent!) would be negligible. On the other hand, if he recognized that, Swift could not have described (assuming his traveler needed food supply equal to that of 1,728 Lilliputians) either how Gulliver "had three hundred cooks to dress my victuals, in little convenient huts built about my house," nor how he "took up twenty waiters in my hand, and placed them on the table: a hundred more attended below on the ground"—nor how Flimnap, a treacherous minister, informs the Lilliputian king that Gulliver "cost his majesty above a million and a half of sprugs (their greatest gold coin) . . . and, upon the whole, that it would be advisable in the emperor to take the first fair occasion of dismissing" him.

Gulliver's daily routines, from walking to urinating, would continue to pose dangers to Lilliput's tiny inhabitants, who might find themselves too close to his boots or his stream, but he could be supported as a permanent resident of the diminutive realm without any strain on the king's treasury—especially as he would more than earn his keep by providing invaluable size-dependent services to keep in check the kingdom's Blefuscudian enemies. But properly scaled Lilliputians would not be such attentive company. At 80 kilocalories per day, Lilliputians would have to consume more than four times as much food energy relative to their body mass than their giant guest, and hence they would have to devote comparatively more time to procuring and consuming all this extra food.

Of course, if Gulliver were allowed to stay in Lilliput, he might have postponed his departure—and that delay could have precluded Gulliver's adventures in the land of giants, where the proportionality of sizes is reversed. In these upended circumstances, Gulliver's requirements remain the same: resting in a portable bed-chamber box made by a Brobdingnagian cabinet-maker and taking a short stroll, he would need no more than 2,500 kcal/day; Gulliver entertaining the king and queen by such feats as running along a spinet-like instrument nearly 60 feet long and banging it with a cudgel-like stick in order to play an English tune for the royalty would need around 3,000 kcal/day.

Because the 3/4 metabolic exponent appears to work reasonably well across the entire span of the mammalian class, we can apply the rule to Brobdingnagians: the BMR of an adult male (weighing 9.7 tons) would amount to about 68,400 kcal/day, approximately 27 times Gulliver's usual rate. Again, that would be much less than would result from scaling food needs as the third power of that gigantic mass. A metabolic rate significantly lower than that implied by Swift has many inevitable implications, and an interesting example illustrates the effect.

Swift relates how, despite her "weak stomach," the Brobdingnagian queen "took up, at one mouthful, as much as a dozen English farmers could eat at a meal, which to me was for some time a very nauseous sight." A bit of algebra will dismantle that claim. Let us assume that an early 18th-century English farmer—necessarily a fairly big eater, given his daily work in the fields and with animals—would need at least 3,000 kcal/day, or 1,000 kcal/meal. Twelve farmers would then consume 12,000 kilocalories at a meal, and to make Swift's equivalent work it would mean that the queen would be eating no more than six such mouthfuls a day (68,400 / 12,000 = 5.7), or just two per meal. Properly calculated, the queen's mouthfuls, though still impressive, would have been less nauseous—and Brobdingnag's farmers would have had a much easier task to feed this population of properly scaled giants, because their food needs would not go up with the cube of their body mass.

All of this raises several obvious questions. How do we explain the 3/4 rule, and how universal is it in nature? Does it apply to all animals, or is it valid only for mammals? Is the insistence on the widespread validity of the 3/4 law akin to "seeing" the golden ratio everywhere? And if the 3/4 rule offers a reasonable quantification of human food needs, is it possible to work out our own, specific food needs based only on our weight?

Metabolic theories, exceptions, uncertainties

Why should power laws in general, and 3/4 scaling in particular, come up so frequently when we study the relative sizes of parts (or functions) of organisms whose body masses span many orders of magnitude? Since the 1970s there have been many attempts to explain the origins of 3/4 scaling, but after nearly half a century of efforts there is still no universally acceptable causal explanation. The most sweeping explanation, valid also for plants, is offered in a series of publications by three American scientists: physicist Geoffrey West and biologists James Brown and Brian Enquist.[8]

They describe a generic organization of living organisms as networks of tubes that evolved to distribute resources and remove waste. In humans and animals these networks consist of arteries, veins, and capillaries that circulate blood to transport oxygen, nutrients, and hormones and lymphatic fluid to remove toxins and waste. Plants have two kinds of vessels: xylem, which moves water and the minerals

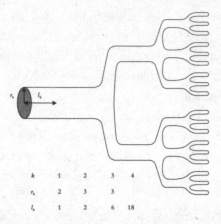

This simplified graph shows just three levels of branching,
from aorta to capillaries: V_k is the number of branches originating
from each conduit at level k; N_k is the total number of conduits.

dissolved in it from the roots; and phloem, which distributes the
products of photosynthesis from leaves to the rest of plant. Because
the distribution networks must reach every part of every living organism down to individual cells, their final branches (capillaries) must
have identical size, and the entire delivery process should proceed in a
way that minimizes the overall energy needed for the circulation.

This led them, via a complex mathematical derivation, to conclude that the metabolism of entire organisms must scale with the 3/4
power of their mass, and they proclaimed that they had formulated a
universal scaling law and laid the foundation of a unifying theory of
biological structure and organization of all living species, be they
plants or animals. This has not been universally accepted.[9] Obviously, branching distributive structures must be a factor in metabolic
scaling, but as Canadian biologist Charles-Antoine Darveau and his
colleagues noted, when explaining the work of any complex system
it is always preferable to consider many factors that contribute to its
function (in this instance, metabolic scaling) rather than advocate
single-cause explanations.[10]

Subsequent plant studies found no support for 3/4 power scaling
of their overall metabolism (some ended up strongly supporting isometric scaling—that is, an exponent of about 1).[11] Other physicists

and biologists have argued for simpler explanations of the 3/4 power
law, while new, and more appropriate, scaling studies have shown
significant departures from the metabolic exponent of 0.75. An analysis that included more than 600 mammalian species (whose body
mass spanned five orders of magnitude), excluded non-basal metabolic rates, and took into account differences in body temperature,
ended up with the exponent of 0.675, confirming Rubner's old dog-
based calculation![12]

Rubner's 2/3 exponent was also confirmed for a large sample of
birds, while analyses for individual mammalian orders showed exponents ranging from as low as 0.58 for insectivores to 0.77 for
carnivores.[13] More fundamentally, it might be more revealing to
measure the field metabolic rate (FMR)—the best indicator of
longer-term energy requirements under typical living conditions.
This rate is always significantly higher than basal or resting rates, and
its measurements are possible thanks to an ingenious technique
invented and developed by Nathan Lifson and his colleagues at the
University of Minnesota starting in the late 1940s: we measure the
loss of water that is marked by two heavy isotopes (atoms with different mass but the same chemical properties) of hydrogen (deuterium,
^2H) and oxygen (^{18}O).[14]

As these isotopes leave the body and are replaced with dominant
lighter isotopes (^1H and ^{16}O), we can measure not only the water loss
but also the amount of CO_2 in the expired air and hence the oxygen
(metabolic energy) demand. This method provides a highly accurate
record of overall energy expenditure over a period of hours and days.
Analysis of the field of metabolic rates of more than 200 mammals,
birds, and reptiles using this methodology revealed a range of exponents between 0.59 for marsupials and 0.92 for lizards, and with no
strong support for the 3/4 law.[15]

Moreover, there are major differences among both mammals and
bird classes. Rates for marsupials are about 30 percent lower than for
other mammals; long-distance ocean flyers (albatrosses, storm-petrels,
shearwaters) that spend most of their time aloft have an unusually high
energy demand for their weight, while desert mammals—adapted to
recurrent food and water shortages—have free-living metabolic rates

Terrestrial vertebrates

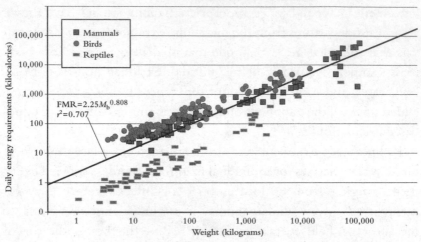

Field metabolic rates for terrestrial vertebrates:
cold-blooded reptiles need less and birds and large mammalian runners
need more than indicated by the average fit line, whose exponent is about $M^{0.8}$.

lower than their body weights would indicate. And specific metabolic rates range widely among animals of the same size but with different modes of living and feeding. Both of these mammals weigh about 3.5 kilograms, but the specific metabolic rate of an Arctic fox, a carnivorous runner, is 2.1 times higher than that of the herbivorous brown-throated sloth.[16]

Yet another study, based on rigorously selected data for reptiles, amphibians, birds, and mammals spanning six orders of magnitude by mass, concluded that the scaling exponents for these four classes of vertebrates are quite different.[17] And, perhaps most convincingly, a meta-analysis of allometric exponents for 127 different species failed to find any support for a universal indicator of metabolic allometry.[18] A range of exponents appears to be a more satisfactory answer than the insistence on a single, universal, all-embracing value. Another way to look at metabolic scaling is to have it, to some extent, both ways.

So far, the most comprehensive study of all kinds of organisms included 3,006 species whose body mass ranges over 20 orders of magnitude—from a tiny bacterium causing infections in rabbits and rodents (weighing 10^{-14} grams) to the Asian elephant (weighing

4×10^6 grams)—and it excluded the possibility of a universal scaling dependence.[19] But it found that mean metabolic rates of major taxonomic groups fit within the range of 0.3–9 watts per kilogram, with a small laboratory mouse near the top of this range and a 4-ton elephant close to the bottom. True, this is a 30-fold difference, but given the enormous diversity of living forms on Earth, whose masses range over 20 orders of magnitude, this is a remarkably narrow range.

As for insects, the most species-rich group of organisms that do not regulate their body temperature, a study of some 400 species concluded that their metabolic rates scale with exponents ranging between 0.7 and 1.0, but another analysis found them to conform to the 0.75 rule.[20] A study of leaf-cutting ants, those busy tropical harvesters who cut and transport leaves to underground chambers in order to grow their fungal gardens, added another interesting insight to the metabolic scaling of social insects.[21] Individual burden (the intensity of foraging) declines only marginally when more workers are added to the workforce, but the trail length limits the foraging rate and hence the amount of resources needed for the colony's growth and reproduction: a trail that is 50 percent longer cuts the foraging rate by half, posing an insurmountable metabolic limit on the colony's size.

As might be expected, size-dependent human metabolic rates also behave in a less than perfectly orderly manner. BMR depends on gender, body size, body composition, and age, and hence it does not account for a fixed share of the BMR but for a range of about 45–70 percent of total daily energy need. Extremes of actual daily energy expenditure based on the latest international recommendations range from less than 2,000 to more than 4,000 kilocalories a day. Total energy expenditure (TEE) is the product of BMR and of the appropriate (and approximate) physical activity level (PAL): TEE = BMR $\times$ PAL (obviously, PAL = TEE / BMR).[22] For a 70-year-old woman weighing 55 kilograms whose largely sedentary life is interrupted by short spells of light domestic activity, the calculation is as follows: a BMR ($0.038M + 2.755$) of 4.85 multiplied by 1.53 equals 7.41 megajoules or 1,770 kcal/day. For a 25-year-old 80-kilogram male lumberjack the total will be BMR $(0.063M + 2.896)$ of 7.93 multiplied by 2.4 equals 19 megajoules or 4,540 kcal/day.

But these population-wide averages cannot be applied with any high confidence to actual individuals. In the late 1940s, Elsie Widdowson—a pioneer of modern nutritional studies and the woman responsible for Britain's wartime food rationing, which left the population in better health than before the war—asked a fundamental question that is yet to be fully answered a lifetime later: why can one person live on half the calories of another, and yet remain a perfectly efficient physical machine?[23]

Commonly high obesity rates present yet another complication: in order to avoid excessively high recommendations of average food energy intakes, obese individuals should not use their actual weight but rather a relatively narrow range of desirable weights when calculating their daily food needs. But the problem goes far beyond individuals: we have known for decades that entire populations use food energy more efficiently than the standard predictions would indicate. In the early 1970s, British nutritional scientist Nicholas Norgan and his colleagues found that pregnant and lactating women in New Guinea did not have higher energy intakes than their non-pregnant and non-lactating counterparts, while the standard expectation is for higher energy intakes under those conditions (during pregnancy, the recommendation is for an additional 360 kcal/day in the second trimester, and 475 kcal/day in the third).[24]

Moreover, daily energy needs among the adults of the coastal Kaul tribe were just 27 percent higher than their BMRs for males and merely 13 percent higher for females (that is, a PAL of only 1.13)—significant departures from expected activity markups of at least 1.5- to two-fold. In the 1980s, a group of British researchers documented lower-than-expected metabolism among pregnant women in Gambia. These women consumed significantly less energy than would be expected even if they were mostly resting, while in reality they were engaged in hard labor every day—and hence further extending their apparent energy shortfalls.[25]

During the 1990s, international comparisons confirmed that standard calculations of basal metabolic rates were significantly over-predicting the values for adults living in the tropics (by as much as about 22 percent for men in Sri Lanka and up to nearly 13 percent for women in India).[26] The verdict is obvious: there is no single value for resting metabolic rate appropriate for all adults; your specific body size is only an approximate guide to your food energy requirements. Calculations using recommended formulas tend to overestimate food energy needs by 10–15 percent even in western populations (whose men and women provided most of the metabolic observations used to derive the predictive formulas), and for many non-western populations (above all, for those in South and East Asia and in sub-Saharan Africa) they might be over by around 20–30 percent.

For biologists—aware of the diversity of life forms and of their complex functions—these uncertainties and irregularities are unsurprising. Capturing life's rich phenomena within the rigid confines of physical laws has proven, repeatedly, elusive. Still, besides satisfying our curiosity about the existence of predictable patterns amid life's complexities, scaling provides many useful, practical lessons. For example, have you ever wondered why we have not domesticated very small animals to produce meat, and why pork is by far the most commonly consumed mammalian meat around the world? Pig weights overlap with our weights; small goats can weigh as little as 10–15 kilograms; small rabbits are just under 1 kilogram, as are the guinea pigs that many Andean families keep in their kitchens, feeding them food scraps and even letting them run free before killing them for their white meat.[27]

If you've read this book closely, you already know why we have not domesticated mice (or rats: they are eaten in many places, but never kept), and you can demonstrate the difference with actual numbers: just deploy Kleiber's equation, using the typical body weights (50 kilograms for a small pig, 10 kilograms for a small goat, 1 kilogram for a rabbit or a guinea pig, 250 grams for a rat, 20 grams for a mouse)—and you will see that even with strictly confined animals, mice would need nearly three times as much feed per unit of their weight than guinea pigs, nearly five times as much as a small goat, and seven times as much as a small pig. On the other hand, cattle are too large and too slow to mature to be rapidly produced in confinement.

Eating, even when resting, seven times as much per unit of body weight as pigs do is no way to mass-scale affordable meat production, especially when the final product (another scaling inevitability!) comes with relatively too much skin. Of course, there are other considerations: a pig's true omnivory (easy to feed, anything goes—while mice prefer grain and do not thrive on liquid waste), its relatively high share of body fat, and hence those desirable qualities of palatability and satiety. Guinea pigs and rabbits are more efficient energy converters than mice, but still too inefficient to make it big in the world of large-scale commercial meat production. Pigs, whose sizes are similar to humans (mostly between 75 and 150 kilograms), have the ideal weight, and nearly 1.5 billion of them are now slaughtered every year around the world (about half of those in China). As always, size matters.

Scaling of artifacts: the metabolism of machines

It is intriguing to note that the same power-law regularities apply to the scaling of man-made objects: to the mechanical equivalents of organisms that "metabolize" (convert) fossil fuels or electricity into kinetic energy. Perhaps the best choice for examining these inanimate scalings is to look at internal combustion engines—the machines that, together with electric motors, are the dominant energy converters and mechanical sustainers of modern civilization.[28] Their two main categories are reciprocating engines with fuel combustion taking place inside pistons, and gas turbines (jet engines) with fuel combustion taking place in a chamber supplied with compressed air.

Mass-scale adoption of gasoline-fueled engines began with affordable automobilization (launched by Ford's Model T in 1908), and soon afterward came the diffusion of more efficient engines invented by Rudolf Diesel—widely used in trucks, locomotives, and heavy shipping. We now have, worldwide, about 1.5 billion road vehicles powered by gasoline- and diesel-fueled internal combustion engines: passenger cars (increasingly as heavier SUVS), pickup trucks, trucks, buses, as well as motorcycles and mopeds. The world's largest diesel engines are powering the world's largest tankers, bulk carriers, and container ships (there are more than 50,000 such ships in service). Stationary combustion engines are mainly water pumps and electricity generators, in off-grid locations and as emergency backups. In total, the world now has close to 2.5 billion gasoline- and diesel-fueled internal combustion engines. Their power ranges from just 5 watts for the world's smallest model plane engine (Tee Dee .010) to 84 megawatts for the largest Wärtsilä marine diesel.[29] This power range, across seven orders of magnitude, is the same as is the body mass range for mammals (shrews to blue whales).

The replacement of engines by more efficient electric motors has been limited by the performance of batteries: even today's best batteries (with about 300 kilowatt-hours per kilogram) have an energy density equal to only about 1/40 of hydrocarbon fuels (>12,000 kWh/kg).[30] The first gas turbines for airplanes were designed during the late 1930s, and their use in military aircraft preceded their rapid adoption in commercial aviation during the 1950s. Turbojet designs were eventually replaced by turbofan engines (with most of the compressed air bypassing the combustion chamber), and before the COVID-19 pandemic more than 20,000 jetliners, passenger and cargo, were in service.[31]

The scaling of internal combustion engines has been studied in detail both for aircraft (fixed-wing and rotary) and for car engines. In the early 1980s, two American scientists, Thomas McMahon and John Tyler Bonner, looked at nearly 40 engines of all kinds (automotive, air, marine) whose power and weight ranged, respectively, from about 330 watts to 21 megawatts and from 135 grams to 102.3 tons, and found that their maximum power output scaled at nearly the same rate as engine mass.[32] A recent study of four-stroke car engines, including Ford, Honda, Kawasaki, and Subaru designs, confirmed this scaling by finding that their peak power output scales with 0.95th power of the engine size.[33] As expected, the scaling of piston aircraft engines is also nearly isometrical. Data for more than 50 engines, from the one powering a 1.5-kilogram micro air vehicle to the four turboprops of the Antonov An-22 (an old Soviet heavy transporter weighing 250 tons) showed that their power is proportional to the 0.9th power of mass and 0.8th power of speed.[34]

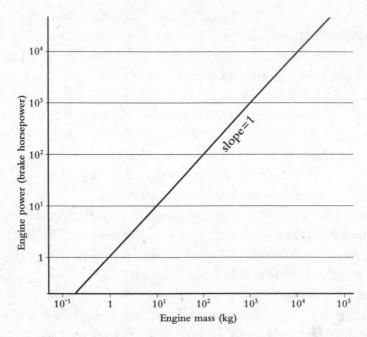

Except for the smallest sizes, the power of internal combustion engines (indicated here as brake horsepower) scales isometrically with their mass.

And I must note one of the most fascinating continuities in the scaling of propulsion. This fundamental property of motors was demonstrated by James Marden and Lee Allen in 2002.[35] They gathered data on the mass and maximum force output of all kinds of motors, interpreting the term "motor" in its widest sense as any organ or a combination of organs enabling motion. Their data set ranged from individual proteins involved in muscle contraction (myosins and kinesins) and flagella (tiny hair-like organelles) that move *Escherichia coli* bacteria, through flying, swimming, and running animals, all the way to piston and jet engines and rockets. Analysis revealed two different scaling regimes for motor force output. In the first group are single molecules, muscle cells, and whole muscles, as well as winches and rising rockets: their maximum force outputs scale as two-thirds power of motor mass ($M^{0.69}$).

The second, much larger group of "motors" includes all organic or mechanical arrangements that move bodies in a more complex fashion than the "motors" of the first category, which simply push or pull loads linearly. Although this second category ranges across ten orders of magnitude—from flying insects and bats, through birds and running and swimming mammals, to electric motors and piston and jet engines—the maximum force output of all of these motors scales at an almost perfectly isometric rate ($M^{1.0}$), with exponents of 1.08 for electric rotary motors and bats, 0.96 for flying birds and aircraft turbines, and 0.95 for running animals!

Isometric scaling makes the multiples simple: doubling the maximum force output requires doubling the motor mass, no matter if the motors are a bird's or bat's muscles, reciprocating engines, or large turbofans. If you want a jet engine whose maximum takeoff thrust is twice as large, you must make it at least twice as massive. In 1970, four Pratt & Whitney JT9D turbofan engines powered the first commercial flight of the first supersize wide-body Boeing 747: each

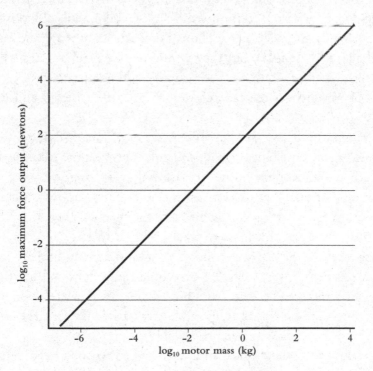

Across all of these sizes the maximum force has
an almost perfect 1:1 correspondence with increasing mass.

engine developed up to 250 kilonewtons of thrust and weighed 4,044 kilograms; since 1996, General Electric's GE 90 has been the world's most powerful turbofan, and its 110B version has a maximum takeoff thrust (512.9 kilonewtons) 2.05 times that of the JT9's—and, at 8,762 kilograms, it weighs 2.16 times as much.[36]

As a result, the thrust/weight ratio of the larger engine (58.5 newtons per kilogram) is almost the same as that of the turbine introduced half a century ago (61.8 newtons per kilogram). Given the 1:1 scaling, this high similarity of specific outputs extends across the entire set of motors belonging to this huge category. Despite their enormous range of body weights, stunning variety of shapes, and widely different modes of locomotion, insects, birds, bats, fish, mammals, electric motors, and piston and jet engines share a key attribute: a remarkably invariant mass-specific net force output, as 1 kilogram of their "motor" mass develops mostly between 43 and 71 newtons, with the mean value about 57 newtons (even the record-size GE 90 turbofan jet engine is very close to this mean). This reality entails obvious size limits. Airplanes twice as capacious as an Airbus A380 could carry nearly 2,000 passengers, but they would require four engines whose thrust and weight would have to double—adding up to nearly 54,000 tons, or as much as the takeoff weight of a fully fueled and fully loaded Boeing 737-600.

The next time you see the white contrail of a high-flying jetliner outlined against the blue sky and, coincidentally—some 11 kilometers beneath it—a flapping goose or a seagull, you might ponder the remarkable coexistence of a rare kinship and the enormous variability of organisms and artifacts: the bird and the machine develop almost identical force per unit mass of their highly disparate "motors." Of course, there is the obvious common denominator: in both of these instances, force outputs have to overcome the weight and inertia of the bodies they are attached to—but the relatively narrow range of maximum specific forces (only about ±25 percent from the mean value) is still a remarkable outcome.

But how does metabolic scaling work when we look beyond mere individuals? Cities are the most obvious entity to investigate as they account for a disproportionately large share of global energy use:

they house about 55 percent of humanity but consume nearly 70 percent of all energies and generate more than 70 percent of greenhouse gases.[37] The new science of urban metabolism has been studying flows of energy and materials in cities, and CO_2 emissions can be seen—much like their human equivalents that result from food digestion—as revealing indicators of metabolic efficiency. Are there expected economies of scale with urban energy use? A study of emissions from US metropolitan areas between 1999 and 2008 found that, contrary to expectations, CO_2 emissions scale proportionally with city size, and that larger cities are not metabolically more efficient than smaller ones.[38] The scaling coefficient was only 7 percent lower than 1.0—that is, every 1 percent rise in population led to a 0.93 percent rise of emissions.

Whether we define metabolism narrowly, as the food requirements of animals, or more widely, as all energy conversion by moving objects, we find that there are commonalities related to underlying functionality. For example, any organism or machine able to fly has to develop a very similar force per unit of mass. At the same time, there are also many departures from expected "rules"—such as animals of the same mass having significantly different metabolic rates—that reflect diverse living conditions and environmental adaptations, and confirm the biologist's old adage: that it is impossible to reduce life's complexity to the mathematical precision of all-encompassing rules deployed by physicists. And so, if the scaling of size and metabolism does not adhere to a single immutable rule, do the distributions of sizes—of living things and artifacts, and of their parts—assume predictable and readily quantifiable forms? We'll see in the next two chapters.

My (more or less systematic) meanderings through the fascinating landscape of all things size have led me from perceptions and illusions to proportions and designs, from basic measurements to the scaling of bodies, organs, and artifacts, and then to the still-disputed regularities of metabolism as a function of size. What remains to be done? So far, I have not said anything about the frequencies with which different sizes appear in nature or among our artifacts. The question remains: how is this enormous variety of sizes—ranging across many magnitudes and belonging to both natural and man-made realms—distributed?

If the sizes of newborns or adults of one species do not resemble each other very closely, to what extent do they differ, and is there any regularity in departures from what appears to be the most common value? Do the wingspans of Amazonian dragonflies, the heights of seven-year-old children in Tokyo, the severity of Californian earthquakes, the populations of French cities, and the sizes of individual fortunes (be they in affluent countries or in the world's poorest regions) show any regular, non-random, and hence predictable distributions that could serve as models to provide foresight and guidance? If they do, then we have a powerful tool allowing us to predict the outcomes of growth or to spot abnormal development.

The most common possibilities are easy to describe and illustrate. Sizes can be distributed in perfectly symmetrical ways around their mean—which in this case is the same as the median (the value separating the data set in two halves) and the mode (the most frequent value)—with more extreme values progressively less abundant and without any distant outliers. They can form asymmetric arrays (where means, medians, and modes do not coincide), skewed either to the left or to the right. They can be bimodal, with two distinct frequency peaks. They can assume two extreme forms: a more-or-less uniform

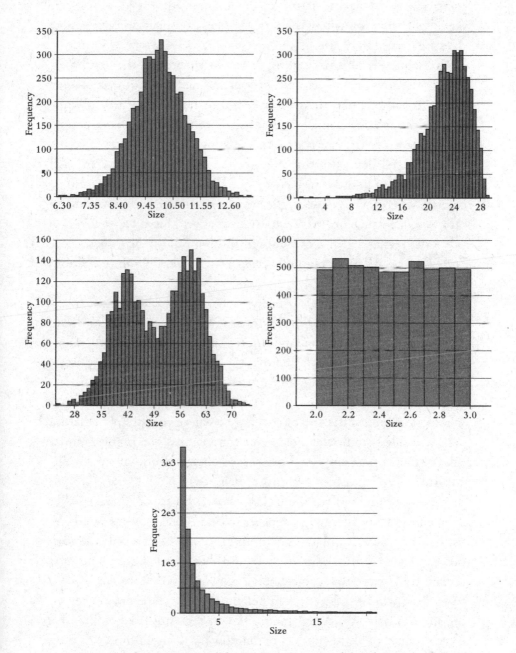

Symmetrical clustering around the mean; asymmetrical distribution (with the mean shifted to the right); bimodal pattern (with two peak frequencies); a highly uniform distribution; extremely skewed distribution with small sizes dominant.

pattern with no size deviating far from the mean, or an extremely skewed distribution where the smallest sizes dominate and the largest sizes are exceedingly rare.

Consequently, it is fascinating to see how common the first possibility is—how many natural properties and phenomena center symmetrically around their mean—and to realize how this ubiquitous attribute has been subconsciously internalized even by people who have never taken a course in statistics and have never encountered the well-defined terms used to describe size distributions. All that is required in order to be aware of the large range of size distributions where values close to the average dominate and extremes are rare, is to have such ordinary experiences as walking on a packed city sidewalk, seeing a group of children spilling out of the doors at the end of a school day, or coming across a sunflower field in full bloom during a summer drive.

The sidewalk experience is easily replicable anywhere in the modern, highly urbanized world: it does not matter if you walk in Milan or Manila, you will immediately notice when an adult man's height strays well above (or stays far below) the dominant range that clusters around 175–180 centimeters in Europe and about 10 centimeters less in East Asia.[1] Similarly, no matter where you see a class of grade-schoolers leaving at the end of the day, or where you admire a mature yellow-headed-sunflower field, you cannot miss the highly similar sizes of these organisms—as both sets of observations will be dominated by individuals of near-identical heights.

Among children of the same class, those who are one head smaller or one head taller than what appears to be a typical stature would be very uncommon. Similarly uncommon would be the sight of sunflower plants rising far above their neighbors, and those plants that remain well below that typical height would be similarly rare.[2] To be sure, there will be differences in height, but only rarely will they go beyond a relatively narrow range. If you measure the heights of a large number of third-graders or sunflowers, you will find not only that most of the departures from the mean are fairly small, they will also cluster fairly symmetrically around their respective means—some a bit lower, others a bit higher, but none shockingly apart from the rest of the set, and none skewed toward any extreme. There is a critical caveat: the size of the group is important. If you wish to present your conclusions with a high level of confidence and a low margin of error, then you will have to measure every item in a small group and make sure that you take an adequate sample of a large group (it would be too tedious, and quite unnecessary, to measure every sunflower in a field).

Any observant person knows that taking just a handful of measurements can give you a very biased result—but how far do we have to go to get reliable numbers? There are many specific quantitative ways to determine a meaningful size for a sample, and there are many sample size calculators available online. As a general rule, for the most reliable results you should measure every item in a group smaller than 100, but if a group contains 1 million items then a sample of less than 400 will do.

Properly measured normal distributions will form variously shaped "bell" curves (one of countless possibilities is shown in the first image of the figure on page 167). These distributions are encountered frequently, registered subconsciously, and because of their commonality they are expected and hence their presence and ubiquity seem normal—and they are actually called so by statisticians. Moreover, this subconscious awareness of normal distributions applies even to rarely seen attributes. For example, only some ornithologists and birdwatchers in North America know the average wingspan of a blue jay in flight, but nearly everybody who has briefly but repeatedly seen this handsome bird would quickly notice if it were double the normal size.[3]

Ultimately, any solid conclusions about the presence of regular distribution patterns of sizes in nature and society rest on rigorous quantitative analyses, but the fundamentals of the subject are intelligible to anybody with just basic numeracy. Remarkable regularities of symmetrical size distributions make it possible to offer either universal, or at least broadly valid, conclusions and predictions in fields ranging from neonatal medicine to tropical ecology—or to tell an apparel manufacturer the relative distribution of shirt or pant sizes to be produced for a specific market. And no less remarkably, the road to this understanding began centuries ago, with the observation of distant celestial objects.

How normal became normal

That elegant trajectory of a perfect, undented, bell-shaped curve tracing a symmetrical continuous probability distribution appears in countless textbooks and writings across scientific disciplines. At the same time, it is necessary to point out that many curves perfectly conforming to the formula used to calculate normal distribution are not at all bell-shaped: some are much narrower and tighter (tongue-shaped, we might say); others are much less elevated, more like gentle mounds than bells in profile.

As for "normal," the qualifier would seem to imply that distributions other than these orderly arrangements around dominant means are abnormal, unusual, or exceptional. This is not true: as already noted, many sizes are asymmetrically distributed—and the resulting curves or (when transformed into logarithms) straight lines are not called "abnormal"! How did we adopt, and then keep on using, this misleading terminology? A brief recounting of relevant astronomical, mathematical, and statistical history will explain how normal became normal. Those smooth rising-peaking-falling curves have a nearly

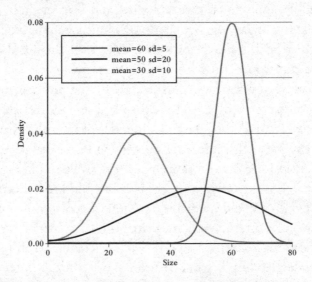

Normal curves with different means and standard deviations.

300-year history, starting with tentative searches for improving the accuracy of observations relying on early telescopes, advancing to their incipient numerical formulations, reaching the stage of clear definition, and eventually moving far beyond their original applications—to the measurements of many organisms and phenomena.[4]

The quest for such clearly defined distributions did not start with ranking sizes, but with trying to compute probabilities and understand the errors in astronomical observations. During the early generations of modern, telescope-aided astronomy, such errors arose due to limited capabilities of observers—as well as due to the various inadequacies of instruments used under less than perfect conditions. Just before 1600, Danish astronomer Tycho Brahe (1541–1601) was the first to favor repeating and combining observations of the same phenomena—commonly the right ascension, the east-west coordinate (similar to terrestrial longitude) used to measure the position of a planet or a star—in order to improve the accuracy of the actual position of a celestial body.[5]

But it was Galileo Galilei (1564–1642)—forever remembered for his *eppur si muove*, and already cited in the fifth chapter in connection with the proper scaling of bodies—who published the first clear summary of the problem. He noted that, due to the observer or the instruments, all observations have errors; that small errors are more common than large errors; and that the observation errors are distributed symmetrically—but he did not suggest how to handle this problem mathematically.[6] The two obvious options were to take either the mean (the sum of the observations divided by the number of observations) or the median (the middle value, with the same number of observations below and above it). The quantification of errors in observations was first systematically examined by the French mathematician Abraham de Moivre (1667–1754). In 1733, after years of work, he published a paper on methods whose applications make it possible to deduce "some practical rules to estimate the Degree of Assent which is to be given to experiments."[7]

This work was eventually recognized as the original appearance of the normal curve. But de Moivre's work was eclipsed by two mathematicians whose names became synonymous with the symmetrical distribution of errors: Pierre-Simon Laplace (1749–1827), whose research contributed not only to mathematics and statistics but also to engineering and philosophy, and Carl Friedrich Gauss (1777–1855), perhaps the greatest mathematician in history. As a result, many statistical—and other scientific—publications used the labels Gauss or Gauss-Laplace (and, in France, Laplace or Laplace-Gauss) distribution.

Laplace advanced its formulation for the first time in 1774 but the distribution became widely known as Gaussian after Carl Friedrich Gauss discussed its properties in 1809 and in his mathematical proof used the Latin term *normalis* (perpendicular, at right angles), which was later misinterpreted as signifying a conventional, "normal" distribution. And soon afterward, Laplace provided an even better mathematical foundation for analyzing the distribution of errors.[8]

The next step was to recognize that the symmetric decline from the mean is also valid for statistical observations of many natural variables on earth. Adolphe Quetelet (1796–1874), Belgian statistician, mathematician, and astronomer, was the first scientist to take that step—when he applied the distribution used to appraise observational errors in astronomy to large-scale anthropometric measurements that had become increasingly available thanks to the emerging field of quantitative social sciences.

Quetelet's 1835 book on "social physics"—translated into English in 1842 as *On Man and the Development of his Faculties, or Essays on Social Physics*—introduced the concept of the average man (*homme moyen*) by examining many physical and mental characteristics of some western European populations.[9] Eleven years later, in his letters to the young dukes of Saxe-Coburg and Gotha (he tutored them previously), he offered the first two examples of normally distributed human attributes based on a large number of measurements.[10] The first set, extracted from the *Edinburgh Medical and Surgical Journal* of 1817, measured the chest girth of 5,738 Scottish soldiers. Quetelet claimed that these measurements were distributed normally—that is, symmetrically around the mean.

Actually, the bar graph of the frequencies was far from being highly perfectly symmetrical, but the fit was close enough to support Quetelet's further inquiries. His second attempt to fit a curve involved the heights of 100,000 French men conscripted in 1817. An excess of men

Pierre-Simon Laplace and Carl Friedrich Gauss provided
the theoretical foundations; Adolphe Quetelet made
the first statistical applications to data sets of public interest.

in the shortest class (less than 1.57 meters) and a deficit in the next class (up to 1.6 meters) did not conform to a normal distribution, but Quetelet, eager to confirm his preconceived ideal, tried to explain this away by noting that the shortest class was exempt from service and hence the surplus of men in this category was evidence of fraud.

Quetelet was impressed by the frequency with which the symmetrical distribution of sizes is encountered in nature, and although many subsequent attempts to fit normal curves to various data sets had mixed results, his work opened the gates for using statistical models to order, appraise, and explain distributions of large data sets. Eventually, the de Moivre-Laplace-Gauss error curve acquired its now-common description as "normal distribution." Three scientists—an American, C. S. Peirce (in 1873); a German, Wilhelm Lexis (in 1879); and an Englishman, Francis Galton (also in 1879)—also independently proposed the term.

Francis Galton (1822–1911), who worked in several scientific disciplines and became famous for his studies of inheritance (and infamous for his advocacy of eugenics, a term he introduced), was the most enthusiastic promoter of the curve's use, and his eloquent appraisal is worth quoting at length, as it conveys his undisguised wonder:

> I know of scarcely anything so apt to impress the imagination as the wonderful form of cosmic order expressed by the "Law of Frequency of Error." This law would have been personified by the Greeks and deified, if they had known of it. It reigns with serenity and in complete self-effacement amidst the wildest confusion. The huger the mob, and the greater the apparent anarchy, the more perfect is its sway. It is the supreme law of Unreason. Whenever a large sample of chaotic elements are taken in hand and marshalled in the order of their magnitude, an unsuspected and most beautiful form of regularity proves to have been latent all along. The tops of the marshalled row form a flowing curve of invariable proportions; and each element, as it is sorted into place, finds, as it were, a pre-ordained niche, accurately adapted to fit it.[11]

Galton even designed an "instrument to illustrate the principle of the Law of Error of Dispersion" that mimicked "in a very pretty way

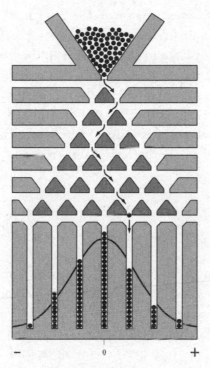

How Galton's quincunx works.

the conditions on which Deviation depends," and had a glass model built by the Tisley & Spiller company.[12] This quincunx device (Galton named it after the arrangement of pins in a geometric pattern— identical to the five-spot on a six-sided die)—led the cascade of small shot darting a step to the left or to the right and ending in successive compartments to approximate the normal curve "however often the experiment is repeated." Larger-scale models of the quincunx device (also called a Galton board or bean machine) have been built for public display (one can be seen at the Boston Museum of Science), and a small version is now available in a sleek modern design.[13] You can build one yourself (using wooden pegs inserted into a board or cardboard, and any kind of small balls) or buy one, and spend hours following its unerring recreation of normal distributions.

During the early decades of the 20th century, the position of symmetrical distribution as the typical, orderly, expected arrangement was strengthened when many of the era's leading statisticians— including Robert E. Chaddock and Raymond Pearl—stressed its importance, particularly in life sciences, in their widely used textbooks.[14] And although not everybody used the term "normal," the distribution was usually treated that way—as it was elevated almost into the category of natural laws. But while the regularities are impressive, the realities of actually plotted size distributions are often less enchanting than the perfect shapes created by the quincunx's ideal dispersions.

Normal distributions, giant trees, IQ, and basketball

While it is undoubtedly true that the distributions of many sizes— whose bar graphs (diagrams of discrete variables) or histograms (bars grouping numbers into ranges) appear mound- or bell-shaped—can be closely approximated as normal curves that follow the memorable symmetric shape, it is also quite common that not all plotted values lie on or near a calculated curve. As already noted, Quetelet's attempts with measuring soldiers and recruits showed that such a fit is commonly far from perfect. So what does perfection look like, how are the values distributed within its confines, and what kind of deviations are encountered in the real world?

To those without any mathematical training, the formula for calculating the normal curve looks forbidding, but understanding the curve's determinants and its properties is not. The curve can be completely defined by just two parameters: the mean (usually denoted by Greek μ) and the variance (the spread of data around the mean). Standard deviation (Greek σ) measures the extent of the curve's variability. These indicators are trivial to calculate. Mean (average) is simply the sum of all observed—be it roughly estimated or accurately measured—values (data points) divided by their number. Standard deviation gets a bit more complicated: it is the square root of variance, and the variance is calculated by subtracting the mean from every individual measurement, squaring the results, adding them up, and dividing the total by the number of observations.

The mean of every normal distribution is also its single-most-commonly encountered value as well as its median (the middle number of a complete series). When the mean value is higher (fruits in a large basket of Jonagold apples will average more than 200 grams; the mean for a basket of Jonathan or Gala apples will be less than 150 grams), the curve will shift rightward along the horizontal axis—but it may or may not change its shape. Standard deviation captures the horizontal spread (width) of a distribution, and hence it changes the shape of the curve. Normal curves with smaller standard deviation will be taller and peakier—spreading less than the curves with a larger σ, which will appear more mound-shaped than bell-shaped.

In order to obtain truly representative distributions, we must work with sufficiently large data samples. How large depends on what level of accuracy we require, and there are rigorous ways to set such parameters. Reducing the advice about the minimum limits to a single number is questionable but it has been done repeatedly, suggesting as few as 20 and more commonly 30 values as the threshold for acceptability. Ideally, we should work with complete sets of data (all newborns in a country in a specific year; all adults in a large city). Given perfect conditions, the area within one standard deviation on each side of the mean contains 68.27 percent of all observed values; two standard deviations subsume 95.45 percent of the measured set; and three deviations entail 99.74 percent of it, which is virtually the entire measured group, with only 0.26 percent, or approximately one of out every 3,850 measurements, in this extreme portion.

This 68-95-99.7 rule could be expressed inversely by saying that 50 percent of all observations fall within 0.674 standard deviations of the mean, 95 percent are within 1.96 standard deviations, and 99 percent fit within 2.576 standard deviations. In the past, statisticians had to rely on detailed printed standard distribution tables to do necessary calculations. Now you can download standard normal tables from the Web or simply check them as needed.[15] These tables use z-scores—calculated by subtracting the mean from a given value and dividing the result by standard deviation—which indicate how many standard deviations a given data point is from the mean (the z-score of the mean is 0).

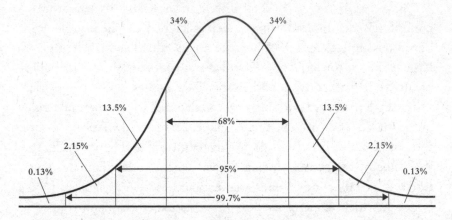

Normal curve, and the shares fitting within specified standard deviations.

Many websites with cumulative normal distribution tables also include interacting graphs, which means you can simply point your arrow and instantly find any probability distribution between the mean and z, up to a given z, or onward from the known z.[16] For example, if an aspiring young tennis player finds that her height is 1.5 standard deviations above the adult female mean (higher stature gives many advantages on court), then she will be taller than 93.32 percent of all adult females (total up to a given z), and shorter than 6.68 percent (onward from the known z).

Understanding standard deviations gives us an excellent tool in appraising how ordinary, or how exceptional, the sizes are that we encounter. In our everyday experiences, our preference is drawn to the mean, and extremes may be both welcome and dreaded. Examples of such expectations abound. We want to see normal, average precipitation, not prolonged droughts or recurrent flood-causing downpours. We want to have the results of all medical tests, from blood pressure to blood cholesterol, within their narrow normal ranges—for example, for high-density lipoprotein, or the "good" cholesterol, it is within 40–60 milligrams per deciliter. And people of average height have a far greater choice and easier task when buying clothes and shoes than people whose size is 6XL—or when, as I will demonstrate later in this chapter, selecting equally sized mates.

In order to maximize their appeal, western politicians address their promises to the average voter—that is, invariably, the not-so-easy-to-define middle-class voter who lives in an average-size house and earns an average-size salary. And parents do not wish to have a very low-weight baby born prematurely at 28 weeks: given modern medical advances, the baby has a very good chance of survival, but its future development may lag behind that of babies delivered at full term (39–40 weeks) with average birth weight.[17] And in terms of height, we know that very few people wish to be much shorter than the average.

Not surprisingly, those relatively rare outliers at either end of the distribution spectrum have often received a disproportionate amount of attention, rewards, or concerns. In nature we admire the world's tallest trees: Douglas firs (*Pseudotsuga menziesii*) growing up to 110 meters and Australian *Eucalyptus regnans* soaring to 125 meters.[18] Deer hunters compete to kill the largest trophy animals with the highest number of antler points, and Josef Fényкövi, an American business-man, gave the largest-ever killed elephant to the Smithsonian after he shot it Angola in 1955: the stuffed beast has been standing in the cen-tral rotunda of the National Museum of Natural History since 1959.[19] At the other end of the natural size spectrum, we marvel at the tiny wingspan of the smallest hummingbird, *Calypte anna*, weighing less than 5 grams, living in Cuba, and making regular appearances on the list of smallest animals (with the blue whale, weighing up to 200 tons, at the top of the ranking).[20]

As for people, in the past some of the smallest children, adolescents, and adults received special treatment at royal courts. In 1626, the Duke and the Duchess of Buckingham presented a seven-year-old and just 45-centimeter-tall (shorter than today's average length for newborn males) Jeffrey Hudson, dressed in a tiny suit of armor, to Queen Henrietta Maria inside a large pie: he then became an addition to the royal collection of extreme bodies, including a giant and other dwarfs.[21] Court dwarfs were so common that many of them have become the subjects of masterly paintings: Velázquez painted individual portraits of five of them (Juan Calabazas, Francisco Lezcano, Don Antonio el Inglés, Don Diego de Acedo, and Sebastián de Morra) and included

Queen Henrietta Maria with Sir Jeffrey Hudson,
painted in 1633 by Anthony van Dyck.

two other dwarfs (Maria Bárbola and Nicola Pertusato) in his most
famous painting, *Las Meninas*, completed in 1656.[22]

During the 17th and 18th centuries, public displays of extreme-size
bodies became common throughout Europe. And during the 19th
century through to the 1920s, cruelly named "freak shows" became a
common form of entertainment on both sides of the Atlantic—
particularly in Victorian England and in the US. Queen Victoria
greatly enjoyed a private performance given by a 62-centimeter-tall
six-year-old called Charles Stratton (advertised by P. T. Barnum as
"General Tom Thumb") in 1844. Dressed in a court suit, he imitated—
as taught by Barnum—Napoleon, and when leaving the room he used
his tiny cane to imitate an attack on the Queen's poodle. Her Majesty
was impressed, writing in her diary that "after dinner we saw the
greatest curiosity I, or indeed anybody, ever saw, viz: a little dwarf."[23]

During the 19th and early 20th centuries, exceptionally tall (and
relatively short-lived) men who joined shows included Angus

MacAskill (2.36 meters), Édouard Beaupré (2.41 meters), and George Auger (2.50 meters).[24] Barnum and Bailey (The Greatest Show on Earth) folded in 2017, but the show goes on, with weight taking over from height—and cable TV replacing traveling displays. Recent American TV shows featuring bodies at the extreme end of the size spectrum have gone beyond ordinary obesity into the morbid territory, and have included *Heavy, Huge, My 600-Lb Life, My Big Fat Fabulous Life,* and *The Biggest Loser.*

And boasting about the size of the human intellect will never go out of fashion. People aspiring to join the ranks of what are supposed to be the world's smartest humans can do so just by taking a test and paying a small fee. Mensa International, registered in England and established in 1946, has many national branches, and in order to hang its ornate certificate of superior intelligence on your wall (or, even better, post a photo of yourself holding it on the Web) you must pass the organization's test—or score at or above the 98th percentile of approved standardized IQ tests designed to have a mean score of 100 and a standard deviation of 15 (with normal distribution, the 98th percentile requires an IQ of at least 131).[25]

Very few people actually take the approved test and pay the required fee, and this limited self-selection keeps the membership small (less than 150,000 worldwide) and conveys a misleading indication of Mensa's extraordinary exclusiveness. But the normal distribution of human intelligence (or, more accurately, of whatever actually gets measured by such questions as "Start with the number of total mittens the numbered kittens lost, and multiply by the voting age in the US. What's the answer?") means that, if the entire nation got tested, more than 6 million Americans would be eligible for Mensa membership, and the global total would reach more than 150 million—a little bit more than Russia's population, a little bit less than all the people in Bangladesh, the world's eighth-largest nation.

One of the most revealing applications of normal distributions is to find out an individual's position on a specific normally shaped curve in order to judge their chances of success (of becoming the CEO of a Fortune 500 company; of marrying a tall supermodel; of belonging to Mensa in Hungary) and therefore their ability to boast of any notable achievements—or to help them come to terms with a nearly guaranteed failure. Unfortunately, too often we do not know the necessary distributions that would allow us to find such placements: it is unlikely that a philatelist in Wuhan could boast about his stamp collection being among the top 10 percent in China (for that, he would have to know the sizes of representative stamp collections in the country).

On the other hand, if you are a young basketball player in Colombia who aspires to play in the US, then all you need to know is your height and have access to the Internet to see what your chances are of ever making it. Because of the rewards and the enormous public interest in these basketball contests, it is not surprising that few distributions of heights are known so accurately (and so persistently). Distributions of heights associated with the lucrative play of sinking balls into baskets are available for both the supposedly amateur National Collegiate Athletic Association (NCAA) and for the professional National Basketball Association (NBA).[26] Statistical analyses confirm the expectation: height is an important factor for the performance of a basketball team.[27] College basketball players obviously come from the right side of the normal height distribution. Given their large number—NCAA's Division I has 353 men's teams, with more than 5,500 players—it is not surprising that the height distribution of this special population subset closely approximates the bell curve whose mean has been shifted significantly to the right of the mean for all American men of comparable ages.

Because the heights are given and plotted in feet and in increments of one inch, I will follow this usage and not convert the values to metric measures. For a quick orientation, and rounded to the nearest centimeter, 6 feet is 183 centimeters and 7 feet is 213 centimeters. The average NCAA height (just under 6'5") is, of course, well above the general population mean (the modal value is even higher, at 6'7"), but the distribution (ranging from 5'5" to 7'6") is almost perfectly normal except for a slightly lower share of men at 5'11" and a higher presence of players 6'7" and 6'8" tall.[28] Also not surprisingly, foreign players (about 15 percent of the total) account for a disproportionately high share of all players taller than 6'9" and for nearly half of all men above 7'0".

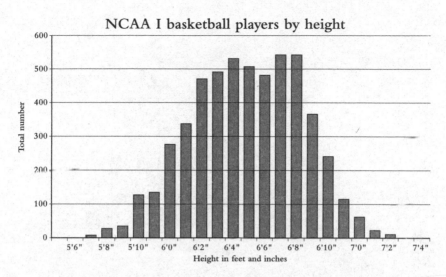

Heights of NCAA I basketball players in the 2016–17 season: men of 6'7"
(200.7 cm) and 6'8" (203.2 cm) rise above the normal curve.

And, yet again, it is not surprising that the height distribution in
the NBA is not nearly as normal as in the NCAA.[29] The total number of NBA players is much smaller (it is less than 1/10 of the NCAA
Division I rosters), and their selection is far more rigorous in order to
justify an average 2022 annual salary of $8.5 million (but the median
salary was "just" $4.37 million) that stands far above baseball, foot-
ball, or hockey league rewards. Given these constraints, the NBA's
height distributions, available since the early 1950s, cannot be
expected to be normal—but the departures from bell curves do not
conform to naive expectations. The NBA does not teem with players
taller than 7 feet (it had only eight of them in 2020) and it does not
exclude men shorter than 6 feet (it had seven of those in 2020).

We have data since 1950 that tells us how individual heights have
contributed to the outcome of games. Looking at the results averaged
over a decade during the 1950s, the height distribution of players was
fairly normal, the modal height was 6'2", and the players of that height
were on court most often—for about 10 percent of all playing time—
while those 6'0" and shorter played 11.4 percent and those 7'0" and
taller just 0.8 percent. Subsequently, the average height increased,
but it peaked in 1987—while the average body mass kept on rising

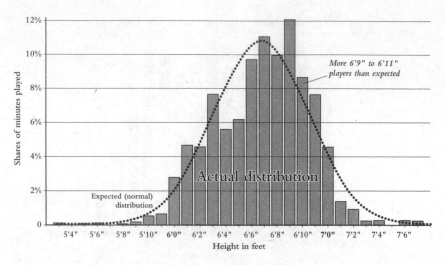

Men 6'9" (205.7 cm) tall have seen more
playing time than their taller or shorter teammates (2010s).

until 2011.[30] During the 1950s, men 6'2" tall led the on-court share of
time (10.2 percent); during the 2010s, the share for the leading height
category hardly changed (10.4 percent) but it now belonged to men
6'9" tall.

Shifts in size distribution are illustrated even more impressively
with the weights of National Football League players.[31] Between
1920 and 1949, their average body mass was, as expected, right of the
general population mean, but it was distributed quite normally: all
kinds of bigger-than-average men could play the game at that level.
During the 1950–1979 period, the distribution of player masses
shifted farther rightward and ceased to be normal: it had a toothy
look, with three similar peaks at 90, 105, and 115 kilograms. And
between 1980 and 2011 the distribution shifted even farther right: it
came to overlap the most extreme range of the general population's
normal curve and became distinctly skewed, with the clear mode at
95 kilograms and subsequent peaks at 110, 120, and 140 kilograms. As
it deviated farther and farther from the general population mean, the
mass distribution of football players became much broader, and its
modes and means became much higher.

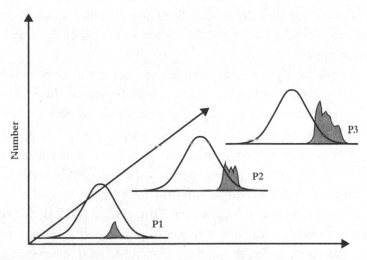

Until 1949 (P1 on the graph) they were just heavier men; during the next three decades American football players became distinctly heavier; by 2011 (P3) they were a group far apart from the general population's mean, with an offensive lineman being close to 320 pounds—about 45 percent heavier than in 1950 and more than 60 percent heavier than the average American adult man.

Normal curves: from antlers to quality control

Once we know the actual mean and the standard deviation of a large set of specific observations, we can answer, with a great deal of confidence, many relevant questions. Some of the most important answers concern human growth, the wellbeing of children, and the worries of their parents. The study of normal distributions gets really useful when it focuses on the extremes.

To be very small is not good for newborns. Infants with low birthweight (less than 2.5 kilograms) have higher perinatal mortality (immediately before birth and up to four weeks after it) and morbidity compared to newborns of normal weight (2.5–4 kilograms); and low birthweight is also associated with higher frequencies of adult illnesses, including cardiovascular diseases, diabetes, asthma, and impaired hearing and vision.[32] Birthweight distribution measured on national scales is almost perfectly normal, with a small departure (2–5 percent of the total) in the lower tail of the curve almost completely accounted for by preterm births. But the mean and standard deviation of the predominant distribution are not infallible indicators of infant mortality.

A notable example are the weights of Mexican American babies: their distribution is shifted leftward (a lower mean) compared to the dominant US average, yet they have a better overall rate of survival during their first year of life (lower infant mortality).[33] Allen Wilcox, an American epidemiologist, had also stressed yet another remarkable finding based on long-term analyses: the average US birthweight has remained—across populations and over time—several hundred grams below optimum birthweight (the weight that is associated with the lowest mortality). This means that, as the average birthweight has risen, so has the optimum weight—and, everything else being unchanged, this shift has had no net effect on infant mortality. Moreover, an international comparison of perinatal mortality showed that Pakistani rates were higher than Norwegian rates regardless of birthweights, and that among the Vietnamese the lowest perinatal mortality coincided with the lowest mean birthweight. Average birthweight thus appears to be a poor predictor of perinatal mortality, and being preoccupied solely with that indicator may not be the best approach to improving the chances of perinatal survival.[34]

Body weights and lengths of newborns cluster tightly, and variations in size naturally increase as children grow. If you think that at 50.5 centimeters your newborn girl is only a bit above average (which is 49.1 centimeters) then you have unrealistic expectations—or, more accurately, a poor understanding of the tightness of the normal distribution of newborn heights. At 50.5 centimeters she is taller than some 70 percent of female newborns, and the height difference between the extremes (5th and 95th percentiles) at birth for girls is only about 7 centimeters; subsequently, it increases to nearly 12 centimeters by the end of the second year and to almost 20 centimeters by the age of 18.[35]

Because of our detailed studies of infants, we can reassure an anxious French mother who is concerned that her three-month-old daughter weighs only 5 kilograms while her friend's child is 1 kilogram heavier. A physician can explain that while the smaller child weighs less than 75 percent of girls in her age group, such a departure from the mean is not in itself any reason for concern. In contrast, because of our understanding of normal growth, a physician must impress upon the parents of a 15-year-old daughter weighing just 40 kilograms the urgency of such a situation: that weight places her more than two standard deviations below the expected norm for her age, and anorexia may be a sobering but likely diagnosis.[36]

At the same time, careful studies have shown that some structural limits have somewhat greater variation than previously believed. The head circumference of healthy newborn babies is one size that cannot depart significantly upward from the average of 35 centimeters. But new research using data from five continents has shown that the standard assumption stemming from the so-called obstetrical dilemma—seeing the size of the birth canal as a tight compromise between having a narrow pelvis that improves the efficiency of movement and a wide, obstetrically sufficient one—is incorrect, and that the geometry of the passage shows remarkable variability, with women in sub-Saharan Africa and Asia showing greater variability than those in North Africa, Europe, and the Americas.[37] But, inevitably, this variability has its limits: a recent American study showed that while about 60 percent of all infants with head circumference equal to or larger than the 95th percentile were delivered vaginally, a large head size is more strongly associated with an unplanned cesarean and instrumental delivery than is high birthweight.[38]

These all are, of course, matters of expert studies, but normal distributions could be used in everyday life for matters ranging from enhancing self-esteem to coming to terms with reality. Unlike beauty (which might be in the eye of the beholder), heights are easily measurable and readily comparable, and the knowledge of normality, mean, and standard deviation offers some powerful tools to assess the likelihood we have as individuals to belong to a particular height category, and to make us aware of expected dominant outcomes for a newborn. As a result, checking some basic anthropometric probability distributions might be either helpful or surprising, disheartening or reassuring.

For a male born in The Hague, the height of 180 centimeters is just below the country's average: Dutch men have the world's highest average height and about 11 percent of them are taller than 190 centimeters. In contrast, among Filipino men (average height of 161.9 centimeters), 99.992 percent will be shorter than 190 centimeters and only one in 13,116 adult men will be taller.[39] Everybody knows that men have higher average heights than women, but how many Americans realize that only about 2 percent of adult American women are taller than the average American man (about 176 centimeters)—but about twice as many men are shorter than the average height for females (nearly 162 centimeters)? Selecting a partner by height may be difficult: if you are a Portuguese male and 180 centimeters tall, you are taller than about four-fifths of all adult males in Portugal. But if you insist on marrying a woman of the same height, then that preference shrinks the pool of compatible partners, because only one out of 265 Portuguese women is that tall—a rather daunting reduction.

And to avoid embarrassment (you never know who might have looked up relevant statistics), you might check the available normal distributions before you start bragging about a pumpkin you grew in your garden for Halloween carving: that seemingly giant squash may be barely of median size among those autumnal cucurbits. The same is true about that large fish you caught (it may only be in the 40th percentile, hence slightly below average) or the deer you shot. Scoring a deer involves measuring the main beam length of its antlers, their inside spread, tine length, and circumference, and adding up the values while also taking into account any growth abnormalities. The score matters to those eager to kill the animals and have their stuffed heads mounted on a wall—or boast about the size.

The record American white-tailed deer scores are more than 300 inches, but such animals are exceedingly rare.[40] Decade-long studies on a private hunting ranch in South Texas confirmed the normal expectation: most buck antlers are average, and very small or really large antlers are rare. Animals were held without supplementary feeding and the mean score was in the low 130s; the range for one standard deviation (68 percent) was 115–151; for two standard deviations 97–169.[41] Knowing this, you can make realistic promises about the size of expected trophies and, obviously, you can ask for an exceptional price when one of your bucks in that ranch herd scores 190.

But there are many more practical uses of normal distribution than scoring deer antlers, and standard deviations have been of enormous help in steering modern industrial production in the desirable direction of high quality and extraordinary reliability. In our modern mass-consumption society, few goals are more imperative than to eliminate—as much as possible—any variability in modern industrial production and to deliver, every year, millions and billions of virtually identical items, be they small wedges of French *La vache qui rit* processed cheese by the Bel Group, now available in more than 100 countries on five continents, or Intel microchips running laptops around the world.

In order to achieve this near-perfect uniformity we need extraordinarily exacting performances: often we must go far beyond three standard deviations when dealing with such specific measures as length, thickness, or diameter of a machined, cast, or 3D-printed part whose defects are normally arranged around the designed mean. In fact, the most desirable level of performance is to keep many of these mass-produced sizes within six standard deviations (6 σ) of a designer's specification. The 6 σ performance goal introduced by Motorola during the late 1980s (the designation "Six Sigma" was patented by the company in 1993) has now been widely accepted—if not as the actual target to be routinely met, then at least as an aspirational goal—by many leading companies worldwide.[42]

Reaching 6 σ performance would mean having only 0.002 defects per million (two parts per billion) opportunities (DPMO). Not surprisingly, such exactitude (a defect rate of a mere 0.0000002 percent) is not easy to achieve, and it is even more difficult to maintain in the long run as the actual performance will drift. This is why a pragmatic approach to the long-term defect rate has taken into account the expected (and often simply unavoidable) deterioration in performance. The actual long-term defect rate is set 1.5 σ lower, at 4.5 σ, and (to confuse the uninitiated) its corresponding DPMO of 3.4 (0.00034 percent) appears as the goal of 6 σ performance in many published tables that do not make it clear that they refer to the performance

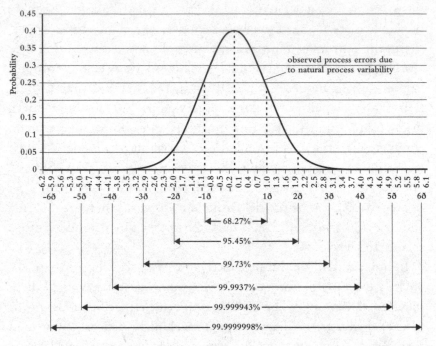

The road to six sigma.

that includes the 1.5 σ shift. Maintaining that level means that a maker of mass-produced parts should achieve a marketable yield of 99.99966 percent—or that a company is able to limit the size of rare material losses, the number of contaminated specimens, or infractions of sanitary regulations to the same near-perfect level.

The importance of minimizing DPMO is obvious for any enterprise engaged in producing billions of units every year: in 2020 Coca-Cola was selling about 1.9 billion servings of its beverages.[43] Other processed foodstuffs with annual giga-sales (more than a billion items sold) include baked goods, dairy products, packaged meat cuts, eggs, snacks, and chocolates. Many things can go wrong along increasingly complex production chains, and food processing requires particularly high levels of vigilance, exercised through constant quality control, in order to prevent improperly mixed batches, poorly cleaned filling machinery, incomplete pasteurization, and the accidental mixing of metal or plastic particles into the finished product.

In order to maintain high levels of consumer approval even at an

otherwise impressive (unshifted) 3 σ level would be quite unacceptable: with a defect rate of 0.2699 percent it would mean that a daily
output of just 1 million units would include about 2,700 defective
cans of soft drinks, or 2,700 only partially pasteurized wedges of
cream cheese, or 2,700 snack bars containing inedible (or possibly
tooth-breaking) matter. Or, taking a competitive perspective, a company producing 100 million mobile phones a year with a shifted 6 σ
(that is, an actual 4.5 σ) would have to throw away just 340 units per
year; its rival able to achieve actual 3 σ would be throwing away
269,900 mobiles every year.

Most people are unaware of the ubiquity of normal distributions
and of their consequences. And yet our lives are influenced, often to a
surprisingly high degree, by where we—and the products we make—
fit along those curves: from the genetic roulette that assigns us
below- or above-average heights (we have seen how many correlations, including major lifelong implications, that position on the
normal curve carries) to the defects per million opportunities in modern industrial mass production (where exceeding a carefully established
rate makes the difference between profit and loss, or between good
and exceptional gains). Size matters, and so does its distribution.

Except in those cases where data availability created an opportunity
to satisfy pure curiosity and to produce, as did the early 20th-century
British preoccupation with all kinds of anthropometric correlations,
some strange findings. The Central Metric Office of Scotland Yard was
storing the measurements of both habitual and less serious male offenders serving their sentences in the major prisons of England and Wales,
and W. R. Macdonell, one of the leading early practitioners of biom-
etric studies, discovered that two prominent sizes—head breadth and
body height—had distributions closely conforming to the normal
curve.[44] A few years later, William Sealy Gosset—chemist, statistician,
and also the Head Brewer of Guinness—sampled the Scotland Yard
data and (writing under the pen name "Student") confirmed that the
size of the left middle fingers among the English and Welsh offenders
was also normally distributed.[45]

From eminently practical and highly profitable to unusual, bizarre,
and rather useless, normal distributions fit them all.

VIII   WHEN ASYMMETRIES RULE

In any symmetrical distribution, the mean coincides with the mode, and knowing the mean—for example, that the average height of 10-year-old Swiss boys is a smidgin below 140 centimeters—can be very useful for many practical reasons, starting with the design of well-fitting clothes and recommendations for adequate nutrition. In contrast, there is no shortage of size distributions that cannot be characterized by their average or typical values: highly asymmetric size distributions have often large, even enormous ranges, from one extreme value that is limited to just a single observation to the other extreme that exists in abundance. Japanese settlements offer a perfect example: at one end, alone above the rest, is Tokyo, the world's largest megacity with a population approaching 40 million people; at the other is a multitude of tiny, depopulating hamlets in Japan's mountainous regions.

Distribution of incomes is similarly skewed, and hence it is obvious that it would not be very useful to calculate the average annual income of 10,011 people living on a tropical island with very high income inequality, where 10,000 poor locals have nearly identical low earnings, clustering tightly around just $1,000 a year; 10 people who own nearly all of the island's economy average a million dollars a year; and one foreigner—who bought the island's largest promontory and built on it a sprawling villa with panoramic windows, fresh and saltwater pools, and a helipad—is making a billion dollars a year. The population's average income will be just above $100,000 a year, a total which nobody actually earns and that is two orders of magnitude (100-fold) higher than the more telling modal income of $1,000 earned by 99.89 percent of the island's inhabitants.

But you do not have to resort to an imaginary island in order to encounter such extremes; you come across them repeatedly by just reading or watching the daily news. Examples are as common as seeing yet another report about a driven mountaineer who has climbed all of the world's peaks taller than 8,000 meters (now it has to be done within some incredibly short period of time to warrant a notice!)[1] or watching a TV program about the Amazon river. The planet has only one Qomolangma (Mount Everest), rising to 8,848 meters; just five mountains taller than 8,200 meters; only 14 peaks taller than 8,000 meters; just over 100 mountains above 7,200 meters—but, as almost any vista in any hilly part of the world demonstrates, there are very large numbers of hills rising modestly above the surrounding countryside, and in flat parts of the world there are innumerable protuberances barely elevated above the surrounding plains. Extremes, not ubiquities, attract: people are not lining up to scale nameless hillocks, but (despite the cost and the mortal peril) they do line up with their Sherpas every spring hoping to make it to the world's most elevated bit of ice and snow.

And there is only one mighty Amazon, with its globally unrivalled flow—much as there is only one Misi-Ziibi ("Great River," its Anishinaabe name) in the US, compared to a huge number of tiny creeks and brooks that feed the tributaries that flow into it.[2] Similarly, there is only one Greenland (an island of more than 2 million square kilometers), just four other islands larger than half a million square kilometers, thousands of tiny uninhabited islands—and every receding tide leaves exposed myriads of temporary mini- and micro-islands made by small rock and sand formations. Why have these distinctly non-normal distributions not made a greater impact on the creators of modern statistics—who, as already explained, were preoccupied with normal distribution? This is even more puzzling given the fact that (as I recounted in the preceding chapter) astronomical observations were at the beginning of statistical theory, and that the apparent size of stars, the most obvious objects in the universe, is distributed in a highly asymmetrical manner. Moreover, asymmetric distribution conforms to a much simpler formula than does the symmetric curve. So what are we to make of this?

The duality of size distributions

Just before the Second World War, Jaromír Korčák (1895–1989)—a Czech statistician, demographer, and one of my professors at the Carolinum University—generalized the great disparity between size distributions that are highly centered and those that are highly skewed when he wrote about "two fundamental types of statistical distribution" (the original in French: *deux types fondamentaux de distribution statistique*).[3] As we have seen, distributions centered on their means are omnipresent in nature—in plants, animals, and humans—and they are encountered in the sizes of entire organisms or their organs and parts as well as their functions (the brains of impala antelopes; grains of wheat harvested in Turkey; heart rates of elite athletes). In contrast, asymmetric distributions in nature prevail where physical (tectonic, geomorphic, atmospheric) forces dominate.

In 1938, Jaromír Korčák
formulated the duality of statistical distribution:
symmetries and asymmetries in nature and in human affairs.

The natural dichotomy of distributions is undeniably ubiquitous, consequential, and relevant, not only from a narrowly statistical point of view but as a problem with obviously broader significance. Its appreciation is useful for understanding not only the entire natural world, but also the outcomes of our intellectual, economic, social, and artistic striving. Inevitably, as is the case with nearly all appealing grand theories, closer inspections show that the duality of statistical distributions —symmetrical versus asymmetrical—is much less clear-cut than a simplistic split between the symmetrical distributions of life's attributes and the asymmetric distributions of the physical world's features and processes.

Korčák attributed the dominance of highly asymmetric size distributions to external factors, and the symmetrical ordering to internal effects. When I first encountered this division in his lectures during the early 1960s, I was impressed both by its simplicity and by its universality. As Korčák stressed, sizes of the planet's observable features are, almost without exception, distributed in an extremely asymmetric way. He illustrated this ubiquity by providing data on the sizes of European islands, lakes, and watersheds, and also on the lengths of rivers and the depths of lakes. As already noted, in global terms the size of river flows is among the most readily appreciated examples, because many people are aware of the enormity of the Amazon's discharge into the Atlantic.

The volume of water flowing into the Atlantic at the river's mouth is five times that of the volume discharged by the Congo, and there are only two other rivers (the Ganga and the Orinoco) whose flow comes close to that of the Congo—but, depending on the definitions chosen to separate rivers from creeks, and creeks from brooks, and brooks from rivulets, there are hundreds of thousands, millions, or hundreds of millions of other streams.[4] The same extreme size asymmetry that applies to the Earth's surfaces is also a property of dynamic geotectonic processes—sudden, violent releases of energy by earthquakes, volcanic eruptions, and tsunamis—as well as of universal phenomena such as the intensity of solar flares and the size of celestial objects striking the Earth.

During the 20th century there were only two earthquakes with a magnitude higher than 9 on the Richter scale (the Chilean Valdivia quake in 1960, at 9.4–9.6; and the Alaskan quake of 1964, at 9.2), but a large number of smaller tremors pose constant risks for all populations living in seismically active coastal regions of the Circum-Pacific "Ring of Fire." People in Japan and California are largely inured to the near-constant occurrence of mini-quakes (clearly recorded by seismographs but hardly noticeable to the inhabitants of seismically active regions), and they have come to accept the relative rarity of truly devastating events. In post-1900 Japan these have included powerful earthquakes in 1923 in Tokyo, 1995 in Kobe, and in 2011, generating an enormous tsunami, in northern Japan.

California has about 10,000 earthquakes every year, but a vast majority of them (magnitude categories 1 and 2) are not felt by its people.[5] There are only about 100 level 3 tremors, which are felt but cause no or very little damage. And since 2010 the state has had only six earthquakes of magnitude 5 and higher (7.1 Ridgecrest, on July 5, 2019, was the largest) that release enough energy to upset or even to terrify people, and that cause appreciable—although often only localized—damage in densely populated areas. I must note that the magnitude numbers are logarithms of actual intensities, and this means that the 7.1 Ridgecrest quake was not 1.42 times stronger but about 130 times more powerful than a rare magnitude 5 event.[6] And during the past 500 years there was no volcanic eruption coming close to the volume of tephra—the fragmented material spewed into the atmosphere—ejected by Tambora (on Indonesia's Sumbawa island) in 1815: it was at least 30 and perhaps up to 50 cubic kilometers.[7]

Asymmetric distributions are common both in nature and in human affairs. Once we move from individual species (whose sizes, organs, and functions often display normal distributions) to complex ecosystems, we encounter plenty of asymmetries—both of sizes (trees more than 100 meters tall are very rare; short blades of grass a few centimeters tall dominate the height distribution of plant stems) and functions: top carnivores exploit much larger areas than do tiny herbivores (lions vs. shrews, eagles vs. sparrows). As for human affairs, the sizes of our bodies and brains (and IQs, whatever they really measure) are normally distributed. But only a clueless inhabitant of another galaxy might expect, after becoming aware of these highly symmetrical distributions within specific age and gender groups, that similar symmetries govern what humans have created, accumulated, and enjoyed. In all of those realms it is extremes, rather than copious averages, that rule.

Any observant Earthling knows this fact, subconsciously but unmistakably, without ever taking any lesson in statistics. In human affairs there are very few mirror-image symmetries centering around dominant means. Asymmetries rule just about everything, from the size of annual incomes[8] to accumulated net wealth (actually a negative value for many indebted people), and from the output of national economies (the US and China being far ahead of the rest) to the ranking of companies (oil production does not have many Exxon- or BP-size enterprises; furniture retailing does not have dozens of IKEA-size competitors) and cities (there is only one Tokyo, whose population is almost as large as that of Canada). And people who have never heard about highly asymmetric size distribution are made aware of it every time they read about the latest ranking of billionaires, because such individuals account for a very tiny fraction of the total population.[9]

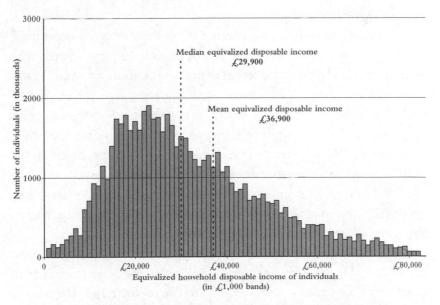

Skewed distribution of average disposable income in the UK in 2020.

Both of these intuitively understood frequency distributions—"normal" symmetries around the dominant mean, and obvious asymmetries of highly skewed size frequencies—are now subjects of detailed analytical and statistical studies, and we have plenty of solid quantitative evidence that allows us to make many useful conclusions, but with different degrees of confidence. In a population with abundant anthropometric data, we can say with a great deal of certainty what will be the most likely weight of a newborn child following the normal pregnancy of a healthy mother. In contrast, as the COVID-19 pandemic demonstrated, we cannot give similarly clear-cut assurances regarding the magnitude of the next major viral infection: will it be a bit more intensive than an annual flu (as was the 2009 event whose death toll was estimated at between 150,000 and 575,000), or will it be much worse than the worst pandemic in modern history at the end of the First World War, which killed at least 50 million people?[10]

Inverse power laws: between rarity and abundance

The prevalence of normal distributions began to be questioned almost as soon as the late 19th century adopted them as a valuable analytical tool. Contrary to Galton's infatuation, the cosmos does not order itself in "normal" ways, but large samples of chaotic elements fit other—asymmetric, but perhaps no less beautiful—distributions. And Galton knew better: although he admired the regularity and the powers of normal distribution, he was actually among those who began to point out how insufficient or outright inapplicable it is to resort to normal distributions: "The ordinary law of Frequency of Error, based on the arithmetic mean, corresponds, no doubt, sufficiently well with the observed facts of vital and social phenomena, to be very serviceable to statisticians, but it is far from satisfying their wants, and it may lead to absurdity when applied to wide deviations."[11]

These wide deviations are hard to miss on the Earth, and the most fundamental (and truly the underlying!) one is the large difference in continental altitudes—from sea level to the top of tallest Himalayan peaks, a range of nearly 9 kilometers. The global distribution of continental elevations is the result of mountain-building (uplift) and water and wind erosion (wearing down) processes driven by plate tectonics and by changing climate, which determines the erosive forces, and it has been unfolding on geological time scales. The distribution is highly asymmetric because continental surfaces are overwhelmingly low-lying.[12]

In the Netherlands, half of the land is less than a meter above sea level (and a staggering 26 percent is below it). Poland's mean elevation (most of it being the North European Plain) is just 150 meters—while in Switzerland half of all land is above 1,100 meters. Switzerland and Nepal (or Tibet and Wyoming) are exceptions, because continental surfaces are mostly monotonous, low-elevation plains. Just over 70 percent of the Earth's dry surface lies below 1 kilometer above sea level; almost 85 percent below 2 kilometers. The Himalayas and other large mountain chains push Asia's mean elevation to about 750 meters, and the averages are 500 meters for North America and 513 meters for Africa (largely because of the relatively high-lying Saharan plateau).

The highly asymmetric distribution that governs continental elevation has had some profound effects on the fortunes of nations— perhaps none as consequential as the repeated invasions of Europe by nomadic populations from Asia's interior. There is no mountain

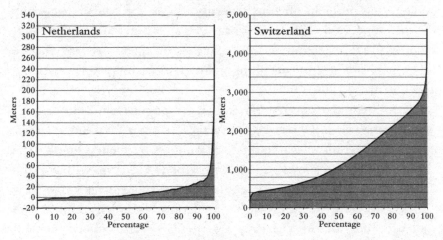

Hypsographic curves for the Netherlands and Switzerland.

range between the Altay range of western Mongolia and the Carpathians of Ukraine and Romania, and the 5,000-kilometer stretch of uninterrupted grasslands served as a perfect conduit for recurrent horse-mounted westward incursions. The history of Eurasia would have been very different in the absence of these extended lowlands. Huns moved in during the middle of the 5th century; Magyars returned repeatedly between 899 and 955 before settling on the trans-Carpathian plain of Alföld, and the most serious threat to Europe came with the Mongol invasions, starting in the early 13th century.[13]

And Eurasia's extensive plains are no exception, as lowlands dominate all inhabited continents. In North America there is no mountain range when traveling northward for more than 4,000 kilometers from the Gulf of Mexico to the Arctic Ocean; and the aptly named Great Plains (and their northern extension, the Canadian Prairies) and a gently sloping basin of the Mississippi west of the Appalachian Mountains fill the center of the continent. This provides enormous opportunities for crop cultivation, but also for extreme weather—as air masses from north and south can travel unimpeded in either direction.

All of the planet's changing physical features (from lakes to sinkholes, from coastlines to streams), as well as the distributions of such dynamic phenomena as earthquakes or cyclones, provide rich areas for analyses attempting to find—as we always try to do—some underlying order in obvious diversity. So do, of course, the many readily observable differences in settlements, income, and wealth. Plotting highly asymmetric distributions on linear axes produces a characteristic, steeply declining curve with a long tail extending rightward.

If you would assign the vertical axis to population totals of individual cities within a large country and the horizontal one for their ranks, you would get a precipitously falling frequency graph. The figure on page 201 shows its actual form for all US cities with populations of more than 200,000 people in 2019: there is only one New York City but 50 cities with populations between 200,000 and 300,000, and (not shown) nearly 700 cities larger than 50,000 people.[14] Obviously, these distributions cannot be well characterized by either

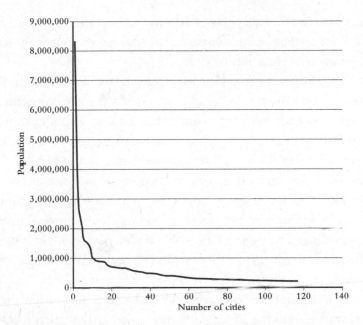

Asymmetries rule the distributions of many man-made phenomena,
including the size of city populations. This graph, using linear scales,
is for all US cities with populations of more than 200,000 in 2019.

average or modal values. Given the dominance of Paris and Tokyo
(Paris houses nearly 20 percent of France's population; Tokyo 30 percent of Japan's) on one hand, and the increasing number of
depopulating small towns and villages on the other, it makes little
sense to talk about an average-size French or Japanese settlement.

In all instances where large sizes (be they devastating magnitude
8 and 9 earthquakes, cities with more than 20 million people, or
personal wealth measured in tens of billions of dollars) are exceedingly rare but small sizes (imperceptible tremors, tiny hamlets,
hundreds of millions of people living below even a draconian definition of the poverty line) are very common, the easiest—if not the
best—way to capture such asymmetric distributions is known as
the power-law function. The basic equation looks far less forbidding than the expression defining normal distribution, and inverse
power-law distributions have the same interesting property as the
positive power scaling introduced in the previous chapter: when

the numbers are plotted on logarithmic (rather than on linear) axes, they fall on a straight line that—rather than ascending from left to right—descends from left to right.

I have already introduced the basic power-law function in the fifth chapter's section on the scaling of animal body organs as functions of body mass ($y = cM^b$, with c as a constant and b as an exponent). All we have to do is to generalize the relationship (x instead of M as an independent variable) and make the exponent negative, to get $y = cx^{-b}$. And, once again, the plotting of inverse power-law functions on logarithmic axes results in straight lines, but while the log-log graphs of organ and metabolic scaling were ascending, these power-law plots are descending. The figures below illustrate a perfect inverse power-law distribution for just five items plotted on linear and logarithmic axes: the straight line in the second graph has a −1 slope.

The first documented observation of an inverse power law did not arise from the ranking of city sizes or plotting the distribution of individual wealth, but from paying attention to a mundane detail. In 1881, Simon Newcomb, a Canadian-born American astronomer and mathematician, noticed that the first pages of logarithmic tables repeatedly used for complex calculations (handheld electronic scientific calculators only became available almost a century later) showed

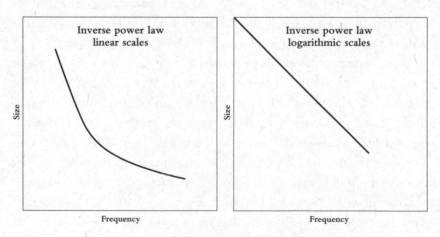

Plotting power-law functions on log-log scales
turns steeply declining curves into straight lines.

much more wear than did the last pages—and he concluded that "the first significant figure is oftener 1 than any other digit, and the frequency diminishes up to 9."[15] Nearly six decades later, Frank Benford, a physicist at General Electric, analyzed a large number of numerical observations, and found that 1 had a frequency of 30.6 percent, 2 had 18.5 percent, and 9 just 4.7 percent.[16] The Newcomb-Benford law thus deserves historical primacy when tracing actual demonstrations of skewed distributions, but perhaps the most widely known power-law distribution came 15 years after Newcomb.

During the 1890s, Vilfredo Pareto, an Italian economist and sociologist, studied tax and wealth distributions in the UK, Prussia, Saxony, Paris, Basel, several Italian cities, and Peru, as one of the key topics of his broad inquiry into political economics. All of his data showed considerable income inequality, but its degree differed substantially among the studied jurisdictions, ranging about ±20 percent around the mean value of his 17 data sets. That is why Pareto carefully avoided giving any single specific measure of this inequality, and instead concluded that: "These results are very remarkable. It is absolutely impossible to admit that they are due only to chance. There is certainly a cause that produces the tendency of incomes to be arranged according to a certain curve. The shape of this curve appears to depend only slightly on the different economic conditions of the countries considered."[17]

In 1913, Felix Auerbach, a German physicist, was the first author to note asymmetric concentrations of populations, and in 1916, Jean-Baptiste Estoup, a French stenographer, analyzed the frequency of French word usage in writings.[18] In 1925, Udny Yule, a British statistician, found a well-fitting power-law frequency distribution of plant sizes in the family of leguminous plants and the families of cerambycid and chrysomelid beetles, and a year later Alfred Lotka, an American mathematician, applied power laws to the frequency of scientific publications within a specific field.[19]

Other pre–Second World War and early 1940s applications of power-law functions included Korčák's already-noted study about the duality of statistical distribution (he analyzed areas and depths of lakes, sizes of islands, areas of watersheds, and lengths of rivers)[20] and

American and Japanese investigations of earthquake intensities. In the early 1930s, Kiyoo Wadati was the first geoscientist to recognize power-law distribution for the energy of earthquakes; in 1939, Ishimoto and Iida applied power law to the distribution of earthquake amplitudes recorded by a microseismograph; and soon afterward, Gutenberg and Richter related the numbers of earthquakes to their magnitude.[21] After the Second World War, Lewis F. Richardson suggested that a power law links the frequency of fatal conflicts with their magnitude.[22] But the most consequential contribution during the 1940s came from George Kingsley Zipf.

This American linguist built on Estoup's observations: during the early 1930s he found that the frequency of using each word in English is almost inversely proportional to its rank in the frequency table. The most common word—"the"—accounts for 7 percent of all occurrences—that is, about twice as frequently as "of" (accounting for about 3.5 percent of all words), etc.—but the rule breaks down after the first 1,000 words.[23] This was one of the first clear hints at frequent irregularities in supposedly regular distributions. Zipf's name became famous only after 1949 thanks to the publication of his book *Human Behavior and the Principle of Least Effort*.[24] Zipf found that decennial census data show the largest city to be roughly twice the size of the second-largest city and about three times the size of the third-largest one—or simply $x = r^{-1}$ where x is city size and r is city rank.

This ranking became widely known as Zipf's law—but if we are to recognize the formulation of a general concept (power laws in asymmetric size distributions) it could also be Benford's law—or Pareto's, Auerbach's, Yule's, Lotka's, Korčák's, Ishimoto's, Gutenberg's, or Richardson's: all of them used the same analytic and explanatory approach. Since the 1950s, power laws (or, more accurately, power-law-like probability distributions) have been used to study the sizes of many natural and man-made phenomena. Besides lakes, rivers, and earthquakes, geoscientists have used power laws to study the sizes of faults, sedimentary layers, oil reservoirs, sinkholes, rockfalls, and volcanic eruptions.[25]

Atmospheric studies have looked at total precipitation generated by rain clusters and the power dissipation of tropical cyclones;

environmentalists have ranked the sizes of wildfires (by areas burned); and astronomers have studied the sizes of fireballs disintegrating in the atmosphere (measured by their impact energy) and the sizes of solar flares.[26] Negative exponents for these size distributions vary widely.

Following Zipf, many studies have ranked city sizes both within their most restrictive administrative boundaries and as sprawling metropolitan areas and conurbations.[27] Economists have looked at company sizes on both national and global scales, ranked capacities of steel plants, and plotted occupational data (the number of people in specific jobs), the length of trips, and income and wealth distributions.[28]

And after the Second World War, Pareto's admonition based on careful observations—"of course, when it comes to purely empirical laws we cannot be too careful"—was ignored, and the ubiquity of asymmetric distributions was popularized in a misleadingly simplistic manner. Joseph Juran, an engineer and a leading American promoter of quality control, found that some 80 percent of all problems with product quality stemmed from about 20 percent of production defects, and this led him to generalize Pareto's observations concerning the uneven distribution of wealth and taxation (a minority accounts for the bulk of the outcome) into a catchy 80/20 rule (and its inverse).[29] In 1997, Richard Koch, a British management consultant, published a bestseller entitled *The 80/20 Principle: The Secret of Achieving More with Less* (with five derivative volumes following between 2002 and 2020), and the 80/20 rule was widely accepted as a universal fact supported by scores of claims referring to a wide range of events.[30]

These have included the sizes of sales (80 percent of them coming from 20 percent of customers) and theft (20 percent of thieves making off with 80 percent of loot). I am not sure how we could—short of measuring sizes of all loots and attributing them correctly to elusive criminals—verify the loot-share rule, but the 20/80 rule certainly does not apply (as has been often maintained) to many sales. Researchers at the Ehrenberg-Bass Institute first reported that, in a year, a brand's top 20 percent consumers (heavy buyers) accounted for an average of 59 percent of all sales, with shares ranging from 44 percent (for hair conditioner) to 68 percent (for dog food), and they later added 15 other results averaging 57 percent for the top 20 percent, with the range from 40 to 79 percent.[31] Clearly, there is a law-like pattern and sale sizes are skewed, but not as much as the Pareto rule has it—the ratio is more like 60/20 rather than 80/20: 20 percent of consumers generate 60 percent of sales, while the bottom 80 percent account for 40 percent of all profits, a share not to be neglected!

And inevitably, in the age of mass-scale communication, power laws have been used to analyze Internet traffic, looking at such indicators as the number of outgoing links from URL documents or the sizes of e-mail address books.[32] Less common applications have included the historic intensities of wars (measured by casualties), the severity of modern terrorist attacks, and the numbers of species in mammalian genera.[33] Given the range of the phenomena whose size and frequencies are claimed to conform to inverse power laws, it is inevitable to ask how close these fits really are: can the vast realm of asymmetric distribution be reduced to a single simple mathematical function?

Orderly asymmetries or wishful thinking?

We cannot expect that most of the real-world data will be perfectly encompassed by simple inverse power-law equations. The efficacy of power laws to capture real-world data is limited, with only a few instances corresponding to what Richard Perline, an American statistician, called "strong inverse power laws."[34] In most instances we have either weak or false inverse power laws, with the expected distribution (straight line on a double-log graph) valid only for a part of the entire data set. In such cases, a specific power law may be applicable only to a truncated segment of the entire distribution, or another mathematical formula may produce a better fit.

City rankings have been perhaps the most frequently studied power-law distributions—and the fit for settlement sizes may be best on the global level, for densely populated areas delineated by the intensity of lights recorded by nighttime satellite images.[35] About 30,000 settlements were identified in this way, and their global

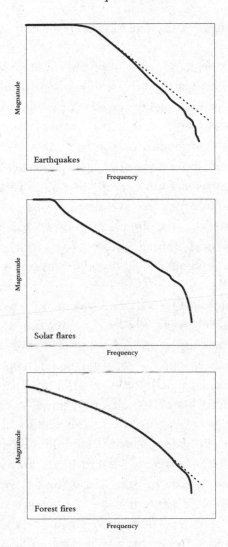

Sizes of earthquakes, solar flares, and forest fires show only partial (truncated) adherence to inverse power law, as straight lines bend—sometimes at both distribution ends.

ranking appears to be remarkably close to a power law. In contrast, even on the global level, ranking metropolitan areas by their population does not follow the inverse power law at the very top: Tokyo's metropolitan area has more than 37 million, the second agglomeration should be less than 20 million, the third one about 13 million—but in

2022 there were eight cities (five in Asia) with more than 20 million people.

We now have several extensive empirical studies testing the power-law hypothesis quantitatively by examining scores of data sets ranging from the sizes of earthquakes to the sizes of meteorites and from the sizes of religious denominations to the sizes of individual wealth.[36] There is no single unique function that could describe all of these distributions, and only a minority of them—including the sizes of US electricity blackouts, meteorite fireballs, and California earthquakes—show a close adherence to power-law distributions. This is how science should work: observing the complexities of the real world, measuring them as best as possible (in this case, as to their size and frequency), and then trying to find out if their distributions conform to an identifiable pattern.

This is how Adolphe Quetelet, building on the work of French and German mathematicians (de Moivre, Laplace, Gauss) introduced the notion of the normal distribution that was eventually accepted as one of the basic statistical models, with widespread applications to size distributions. Not all normal-looking distributions pass statistical tests of perfect symmetry, but even moderately good fits offer useful tools for drawing fairly accurate conclusions and making useful inferences, be they about the expected milestones of childhood growth or about achievable goals for fault-free manufacturing performance.

And the same is true about sizes with strongly asymmetric distributions. Vilfedo Pareto's rule for the size of incomes and wealth, Charles Richter and Beno Gutenberg's rule for the size of earthquakes, and George Zipf's rule for the size of cities became the bywords in, respectively, economics, seismology, and urban studies—even though, on closer examination, the quasi-magical property of power laws that produces perfectly straight lines in doubly logarithmic plots is a mathematical ideal that is rarely encountered in full among real-life distributions. As Mark Buchanan, an American physicist, noted in his review of power-law claims: "wishful thinking can easily slip in, aided no doubt by the seemingly self-evident legitimacy of the reasoning involved . . . the human mind can easily be drawn into incorrect conclusions."[37]

But none of this detracts from deriving many useful insights based on the broader reality of asymmetric size distributions, no matter which function (power law, log normal, stretched exponential) or combination of functions best captures their entire course. Recognizing the reality and ubiquity of asymmetric distributions gives us a fundamentally correct understanding of such complex realities as the probability of extraordinarily massive volcanic eruptions, and of such key determinants of social structure as the wealth accumulated by individuals and families—and hence also powerful insights into the divisions that have characterized all human societies, and the risks faced by global civilization.

That the sizes of many phenomena are not distributed normally has enormous implications for the stability of human societies and for their physical survival. Dealing with the extremes of normally distributed sizes might be sometimes challenging but hardly alarming. The smallest adults are simply the size of children, and achondroplasia (the disorder that is the most common cause of dwarfism) limits body height but it usually leaves intellect or expected lifespan normal.[38] The tallest adults have to be mindful of door heights but they may have a career as a basketballer—and in either case there is no problem providing them with food, clothes, furniture, or medical care. The sizes of morbidly obese patients (twice as heavy as the mean adult body mass) pose challenges for ambulances and hospitals, but we have come up with special loading, transportation, and manipulation adjustments to take care of these extreme body sizes.

In contrast, the extremes encountered with highly asymmetrical size distributions differ commonly by one or more orders of magnitude—and in some cases involving monies they sink into values below zero, making order-of-magnitude comparisons irrelevant. In 2020 the global wealth edifice consisted of nearly 3 billion people (55 percent of the world's adults) whose wealth amounted to less than $10,000; the next category (1.7 billion, a third of the adult population) had between $10,000 and $100,000; only about 11 percent of adults had wealth up to $1 million; and just 56 million (a mere 1.1 percent of all adults) owned more than a million dollars. The asymmetry is shown even more starkly by noting that 1.1 percent of adults

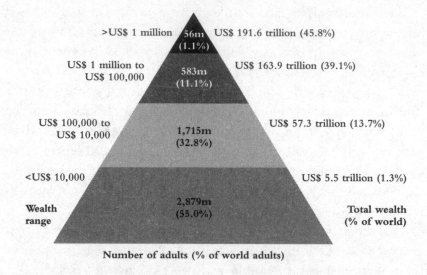

>US$ 1 million — 56m (1.1%) — US$ 191.6 trillion (45.8%)

US$ 1 million to US$ 100,000 — 583m (11.1%) — US$ 163.9 trillion (39.1%)

US$ 100,000 to US$ 10,000 — 1,715m (32.8%) — US$ 57.3 trillion (13.7%)

<US$ 10,000 — US$ 5.5 trillion (1.3%)

Wealth range — 2,879m (55.0%) — Total wealth (% of world)

Number of adults (% of world adults)

An extreme asymmetry: 55 percent of all adults own just 1.3 percent of the global wealth; just over 1 percent of adults control nearly half of it.

command about 45 percent of global wealth, while the bottom 55 percent claim just 1.3 percent of the world's riches.[39]

Differences among nations are substantial. In 2019 the difference between the averages of the bottom 10 percent and the top 10 percent was nearly seven-fold in France but more than 12-fold in the US.[40] Fortunes of a few of the richest individuals in the US are valued in hundreds of billions of dollars (10^{11})—that is, six orders of magnitude (a million-fold) higher than the average net worth of about $122,000 ($10^5$). Certainly the most stunning outcome is that the poorer half of the US population owns merely 1 percent of the country's wealth, and a large share of that group is in debt and has no net worth. Essentially, this makes the extreme wealth differences unimaginably large.[41]

Pareto would not be surprised by any of these realities: his view of history was dark and pessimistic. In his *Cours d'économie politique* he stressed that "we often talk about the social pyramid, of which the poor form the base, the rich the top. To tell the truth, it is not a pyramid but rather a body . . . having the shape of the point of an arrow—or, if one prefers, of the point of a spinning top."[42] And, obviously, the top keeps on spinning, and it does so everywhere and right to its thinnest point: economic inequality in Communist China and billionaires in America offer perfect illustrations of these realities.

China, ruled since 1949 by a Communist Party, styles itself as a paragon of an egalitarian socialist economy, in opposition to what its leaders label as rapacious American capitalism—and yet the country has as unequal a distribution of income as does the United States. The Gini coefficient is the most common measure of economic inequality, ranging from 0 for perfect sharing up to 1, with levels close to 0.5 being undesirably high. China's Gini coefficient has declined slightly from the high 0.49 in 2008 to 0.47 by 2019, but it is no better than the latest US value (0.469 in 2020), and it ranks far above the inequality range for disposable income in the European Union (just 0.3).[43]

Another remarkable reality is that the index has not shown any major decline in either of the world's two largest economies. American economic inequality has been worsening steadily since 1970, when it was as low as 0.36 and the richest 20 percent received 41.5 percent of all income—compared to 50.8 percent in 2020.[44] China's Gini index rose from somewhere between 0.25 and 0.3 in 1980: that rate did not reflect (as does a similar range within today's EU) a more egalitarian division of riches and relative equality of basic affluence, but rather the shared miseries of the Maoist (1949–1979) economy with its food rationing, dearth of consumer products, and curtailed social mobility.[45] And the US example shows how inequalities persist right to the very apex of the wealthiest elite. Analysis of the largest personal fortunes in the US shows that in 2020, among the 50 richest individuals, the top 10 (20 percent) commanded 52 percent of the group's wealth, and among the richest 10 just two men claimed 35 percent.[46]

Inequalities have long preceded the rise of cities—to use just a single notable example, not everybody in Iron Age Europe was buried with a horse chariot, harness, and weapons.[47] But because cities have been the dominant sources of diverse and well-rewarded employment and of transformative technical innovation, as well as sites of concentrated consumption with opportunities for display and indulgence in excess, they should be seen as both generators and products of economic inequalities. As a result, urbanization has been an unstoppable global trend, and everywhere it produces pronounced asymmetries of urban riches and city sizes. Once again, China offers a perfect example. Maoist policies, with food ration tickets redeemable only in the places of permanent residence, severely limited the normal progress of urbanization, and by 1980 only 19 percent (187 million) of the country's 987 million people lived in cities.[48]

Deng's economic reforms resulted in the world's fastest and most extensive urbanization wave in history, as the country made up for three decades of restrictions: by 2020, 64 percent of China's 1.41 billion people (just over 900 million) lived in cities. Shanghai is the world's third-largest megacity and Beijing comes eighth.[49] There may be many good reasons to reverse the growth of cities—and even more so the expansion of megacities—but as long as a national population is increasing, attaining that in a normal society is impossible. Only an exceedingly draconian dictatorial regime might achieve it: we have the tragic example of the murderous, de-urbanizing Khmer Rouge in Cambodia in the late 1970s, whose leaders forced the population to leave the cities and to perish in large numbers in the countryside.[50]

Determined efforts of an interventionist government can, through taxation and subsidies, reduce some of the worst instances of economic inequality.[51] Asymmetric size distributions of catastrophic natural disasters signal different challenges. All too obviously, there is no way to prevent volcanic mega-eruptions, once-in-a-millennium floods, or an unexpected encounter with a massive asteroid impacting the Earth. All we can do is estimate the probabilities of these extremes based on past experiences, but those exercises provide unhelpfully divergent answers depending on how far back our understanding of specific catastrophes goes. During the 100 years preceding 2020, the world's largest volcanic eruption with a clearly discernible global climate impact was Mount Pinatubo in 1991, but it released only about 5 cubic kilometers of tephra. Krakatau in 1883 released 20 cubic kilometers, Crater Lake (7,600 years ago) 150 cubic kilometers, and Toba (74,000 years ago) about 2,800 cubic kilometers.[52] Which category is due for a repeat in the next 100 years?

The latest pandemic has illustrated how fundamentally unanswerable such questions are, even in the cases of much greater underlying understanding. In the pandemic's case we knew, with a high certainty, that it was coming. Epidemiologists kept on publishing their appeals for better-preparedness, and in a book written in 2007 and published in 2008 I wrote that the next pandemic would arrive before 2021 and that we were "probabilistically speaking, very much inside a high-risk zone."[53] We also knew, based on other relatively well-studied 20th-century pandemics (1968, 1957–58, 1918–20) what range of mortality and social and economic difficulties we might expect, and hence what levels of preparedness we should be considering.[54]

And, indeed, some people said we would be fairly well prepared. But we were not, and COVID-19 proved to be, in so many ways, a pandemic like no other, with many surprises.[55] And the same could be true about the coming Krakatau-like volcanic eruption or about the major disruption, if not destruction, of our electric and electronic system by an unprecedented coronal mass ejection—the enormous expulsion of plasma from the Sun that takes place with unpredictable intensity at unknowable intervals.[56] We know some basic contours of what might be coming, but we are ignorant of actual timing, specific properties, and the full consequences of such events. As the latest pandemic demonstrated, it is highly unlikely that we can become adequately prepared for these damaging events arriving from the extremes of asymmetric size distributions.

We prefer order, we favor recognizable patterns, and we delight in finding underlying commonality and relative simplicity among the variety and complexity of the real world. Normal, symmetrical distributions satisfy this preference, and provide a welcome degree of stability and predictability as we encounter them in innumerable natural phenomena. In contrast, it is far more difficult to fit asymmetric distributions into predictably orderly patterns and to make reliable inferences about the frequency with which the sizes of natural hazards and catastrophic events conform to a quantifiable pattern. Still, there is one certain reward: lessons learned from the quest for this elusive order make for a better appreciation of the precarious nature of our existence as a supposedly sapient civilization evolving on a small planet in one of our galaxy's unremarkable star systems.

IX   SUMMATIONS FOR
THE ELECTRONIC AGE

Anybody expecting a grand synthesis culminating in a small number of conclusions imparting concentrated wisdom about size will be disappointed. Size—a fundamental property of both the material and the abstract worlds—cannot be tackled that way. As we have seen, besides obvious commonalities and surprising rules, its very universality guarantees many unique, unusual, and unexpected outcomes and consequences. There are no perfect, crisp, all-encompassing descriptions, no way to subsume realities within a few mathematical expressions: forcing the notions into narrowly preconceived rules simply does not work.

Instead, I am intrigued by the challenge of distilling at least some key findings of this book into brief summaries fit for the electronic age—and to do so in a sequence of sizes diminishing by an order of magnitude: 1,000 words, 100 words, 10 words, a single word. Here are the results.

1,000

An awareness of size is always with us. We assess and compare sizes, register them swiftly, and use them subconsciously to guide our behavior. Other sizes we ponder deliberately, with pleasure, satisfaction, fear, or envy. We are aware of similarities and disparities in sizes, and we have become accustomed to many standard sizes and notice that something is wrong only if they diverge from expected norms. We are keenly aware of many size ranges between small and large— some things we may prefer small, others do not appear ever to be large enough. In general, our imagination tends toward large, and large sizes (viruses notwithstanding) usually carry the most dangerous connotations and are the greatest source of our fears—but they are also the goals of our personal and collective striving: the evolution of our aspirations has been toward larger sizes.

All senses participate in our perception of size, but sight is obviously dominant. We acquire many expected size norms and size preconceptions, and artists have exploited this reality by surprising us with unexpected interpretations. Canonical perspectives dominate our expected views, but even the oldest civilizations were experimenting with size norms and viewpoints. Modern societies have seen the diffusion of ever-increasing sizes of screens, buildings, and machines—as well as of miniaturized electronic devices. And we also continue to see what is not there. This has serious consequences for witness testimonies, and for people with dysmorphic disorders who cannot see the real size of their bodies. Measurements should—but often they do not—set everything right.

Sizes are always seen in relative terms, and proportions—guided by aesthetic preferences or functional imperatives—determine visual appeal and set performance boundaries. Societies have adopted some key size preferences expressed as size ratios. Paintings and sculptures reflect these ideals and pay attention to symmetry. Symmetries are everywhere: in natural designs and in the earliest human tools, as well as in the most famous monumental structures. Symmetries are also preferred for the sizes of our bodies and our faces. Modern artists made asymmetrical sizes common, and sometimes admirable. The golden ratio (approximately 1.618) has often been claimed to be an aesthetically superior asymmetric division. Actually, it is not a ratio but rather an irrational number, there is nothing inherently superior about it, and, on closer examination, it is not found everywhere.

Monumental architecture aside, natural sizes dominated human designs until after 1800, when we began to execute projects and machines of unprecedented size. The resulting increases in sizes, capacities, and numbers in use must be measured not in simple multiples but in orders of magnitude. But there are many limits to growth in size, and s-curves have been forming. Modern electronics appear to be immune to these limits, as microprocessors, following Moore's law, keep on accommodating larger numbers of components. Though this growth, too, has been slowing, and the miniaturization of size is very close to its physical limits. The most immediate design needs—clothes and furniture—remain personal. Designing for comfort and ease of use is the goal of ergonomics, and given the increasingly sedentary nature of modern societies, seat design assumes special importance, particularly when flying.

Size has many important physical implications, above all those regarding gravity, body-heat loss, mobility, degree of activity, demand for energy, and frequency of feeding. As body mass increases, organs scale isometrically (at the same rate as the body mass), hyperallometrically (faster than 1), and hypoallometrically (exponent smaller than 1). Lung mass and blood volume scale isometrically; hyperallometric scaling is rare (skeletons are subject to it); and hypoallometric scaling is most common: brains, hearts, livers, kidneys, total mass of the digestive tract, and the mass of intestines become a smaller share of total body mass as its size increases.

Metabolic scaling (energy demand at rest) was first thought to scale as 2/3 power of body mass, but by the 1930s the exponent shifted to 3/4. The most fundamental explanation of this scaling— based on the properties and behavior of a network of tubes required to distribute resources and remove waste in every organism—looks convincing, but actual rates for large numbers of organisms show significant departures in either direction. Metabolic scaling has a range of exponents, but given the enormous range of body masses (from bacteria to whales, over 20 orders of magnitude) specific differences are surprisingly restrained. The maximum power of all kinds of "motors" (ranging from tiny insect muscles to jet engines— that is, over 10 orders of magnitude) scales isometrically with their mass.

Body sizes of individual species, as well as the sizes of their constituent parts, have normal distribution: their plots form bell-shaped patterns that cluster around the mean. Every normal distribution is fully characterized by its mean and by its variance measured as standard deviation (sigma), the degree of departure from its central value. In normal distributions the mean coincides with the mode, and about 68 percent of all values fall within one standard deviation and more than 95 percent within two. Once these values are known, it is possible to make useful inferences for any of many normally distributed variables, be it height of newborns or basketball players, size of IQ, or blood cholesterol levels.

Symmetries of size distributions have their counterpart in asymmetries of the natural world: heights of continents above sea level, mountain heights, lake or river sizes, magnitudes of earthquakes or volcanic eruptions. Asymmetries also rule social and economic life, ranging from the size of cities, enterprises, and disposable incomes to accumulated family wealth. These distributions appear to conform to specific inverse power laws, but a closer look indicates that only a few of them fit perfectly. Asymmetric distribution of city sizes makes it much more challenging to moderate numerous intra-national and regional differences. The universal tendency toward asymmetric distribution of income and wealth makes it difficult to reduce persistent economic inequalities. And asymmetric size distributions of natural disasters, coupled with their unpredictable occurrence, means that we will never be prepared for the next mega-earthquake or the next pandemic.

100

Size is a fundamental physical attribute. Our size judgments are constant; size expectations and illusions abound; measurements should decide; body height confers advantage. Preferred proportions enhance size appeal and symmetry is generally in high regard. Modernity has seen an overall tendency toward larger sizes. Ergonomic size designs should be everywhere. Allometric scaling of size, including body organs, has a limited range but it is complex. Metabolic size scaling of organisms has no single exponent. Intraspecific sizes of organisms and their parts have normal distributions, clustering around means. But there are plenty of asymmetrically distributed sizes in both nature and society.

10

Size, an omnipresent scalar, is the measure of all things.

1

Size.

References and Notes

I SIZE AS THE MEASURE OF ALL THINGS

1 J. M. van Ophuijsen, et al., *Protagoras of Abdera: The Man, His Measure* (Leiden: Brill, 2013).

2 J. Locke, *An Essay Concerning Human Understanding* (London: Thomas Bassett, 1690), Book III, Chap. VI, § 29.

3 See also: A. Douglas, " 'In a glass darkly': Swift, Gulliver and the human shape," in F. Boulaire and D. Carey (eds), *Les Voyages de Gulliver* (Caen: Presses universitaires de Caen, 2002), pp. 125–38. For more on shape in philosophy, read Aristotle or Locke or modern ontological debates.

4 Commonly used construction lasers have errors as small as 1.5 millimeters.

5 B. Mandelbrot, "How long is the coast of Britain?: Statistical self-similarity and fractional dimension," *Science* 156 (1967), pp. 636–8.

6 M. Mandelbrot, *Fractals: Form, Chance, and Dimension* (San Francisco: W. H. Freeman, 1977).

7 These problems often begin with the definitions of measured variables: how to define literacy or long-term unemployment?

8 Shadow economy is a global, and thriving, phenomenon: L. Medina and F. Schneider, *Shadow Economies Around the World: What Did We Learn Over the Last 20 Years?* (Washington, DC: International Monetary Fund, 2018).

9 Pandemic-induced temporary shortages have brought back calls for barter in rich economies—but, given the size of these systems, such contributions could never be more than limited.

10 M. P. Taylor, *Purchasing Power Parity and Real Exchange Rates* (London: Routledge, 2016).

11 National differences in voltages and plugs amount to perhaps the most obvious standardization failure in the world dependent on charging

electronic devices: V. Smil, "Voltages, plugs and frequencies," *Spectrum IEEE* (July 2021), pp. 20–1.

12 US Department of Labor, Occupational Safety and Health Administration, "Fixed Stairways," https://www.osha.gov/laws-regs/regulations/standardnumber/1917/1917.120.

13 L. W. Smith and L. W. Wood, *History of Yard Lumber Size Standards* (Madison, WI: Forest Products Laboratory, 1964).

14 Online Labels, "What's the difference between US letter and A4 paper sheets?," updated 18 June 2020, https://uk.onlinelabels.com/articles/difference-between-us-letter-a4-paper-sheets.

15 J.-C. Croizé, *Politique et configuration du logement en France (1900–1980)* (Paris: Sciences de l'Homme et Société, Université Paris Nanterre, 2009); US Census Bureau, "Characteristics of new housing," 2021, https://www.census.gov/construction/chars/highlights.html.

16 W. L. N. Tickell, *Albatrosses* (New Haven, CT: Yale University Press, 2000).

17 ExpertAfrica, "The great wildebeest migration," https://www.expertafrica.com/tanzania/info/serengeti-wildebeest-migration; CompaniesMarketCap.com, "Market capitalization of Alphabet (Google) (GOOG)," https://companiesmarketcap.com/alphabet-google/marketcap [accessed 2022].

18 For example, Toba, the largest volcanic eruption during the past 100,000 years, had a Volcanic Explosivity Index of 8.8 compared to 7 for Tambora (in 1815) and 6 for Krakatau (1883). The eruption imperiled the survival of humans in nearby regions, but humans were much less affected in Africa. See E. I. Smith et al., "Humans thrived in South Africa through the Toba eruption about 74,000 years ago," *Nature* 555 (2018), p. 7697.

19 Human brains reached their present-day variation between 100,000 and 35,000 years ago: S. Neubauer et al., "The evolution of modern human brain shape," *Science Advances* 4/1 (January 2018).

20 Box Office Mojo, "Godzilla vs Kong," https://www.boxofficemojo.com/release/rl1383892481/.

21 N. K. Sandars, trans., *The Epic of Gilgamesh*, https://archive.org/stream/TheEpicofGilgamesh_201606/eog_djvu.txt.

22 Penguin e-edition, Fagles (2002); Alexander Pope's classic translation of the *Odyssey* is available at https://www.gutenberg.org/files/3160/ 3160-h/3160-h.htm.

23 D. E. Donnelly and P. J. Morrison, "Hereditary gigantism—the biblical giant Goliath and his brothers," *Ulster Medical Journal* 83 (2014), pp. 86–8.

24 J. Grimm and W. Grimm, *Kinder- und Haus-Märchen* (Berlin: Realschulbuchhandlung, 1812). English translations are available online.

25 J. Swift, *Gulliver's Travels, or Travels into Several Remote Nations of the World. In Four Parts. By Lemuel Gulliver, First a Surgeon, and then a Captain of Several Ships* (London: Benjamin Motte, 1726); L. Carroll, *Alice's Adventures in Wonderland* (London: Macmillan, 1865).

26 C. W. Lippman, "Certain hallucinations peculiar to migraine," *Journal of Nervous and Mental Disorders* 116 (1952), pp. 346–51; J. Todd, "The syndrome of Alice in Wonderland," *Canadian Medical Association Journal* 72 (1955), pp. 701–4; P. O'Toole and E. J. Modestino, Alice in Wonderland Syndrome: A real life version of Lewis Carroll's novel, *Brain Development* 6 (2017), pp. 470–4.

27 English Heritage, "Building Stonehenge," https://www.english-heritage .org.uk/visit/places/stonehenge/history-and-stories/building-stonehenge [accessed 2022]; Centre des Monuments Nationaux, "Site des Mégalithes de Locmariaquer," https://www.site-megalithique-locmariaquer.fr/en/ [accessed 2021].

28 The wonders included the pyramids at Giza, the hanging gardens of Babylon, the statue of Zeus at Olympia, the temple of Artemis at Ephesus, the mausoleum at Halicarnassus, the colossus of Rhodes and the lighthouse in Alexandria: P. A. Clayton and M. Price, *The Seven Wonders of the Ancient World* (London: Routledge, 1990). Obviously, Lincoln cathedral's spire surpassed only the great pyramid's height; the ancient structure's volume (2.6 million cubic meters) was topped only by modern concrete dams. Hoover Dam on the Colorado (completed in 1936) contains 3.33 million cubic meters of concrete, and many dams in the Americas, Asia, and Africa have volumes in excess of 10 million cubic meters: Global Dam Watch, "Global Reservoir and Dam Database (GRanD)," http://globaldamwatch.org/grand/ [accessed 2022].

29 K. Treister, *Maya Architecture: Temples in the Sky* (Gainesville, FL: University of Florida Press, 2013); J.-P. Protzen, *Inca Architecture and Construction at Ollantaytambo* (Oxford: Oxford University Press, 1992).

30 The size of Brunelleschi's unprecedented cupola retains its fascination, with new books devoted to the achievement: F. D. Prager and G. Scaglia, *Brunelleschi: Studies of His Technology and Inventions* (Cambridge, MA: MIT Press, 1970); G. Fanelli and M. Fanelli, *Brunelleschi's Cupola: Past and Present of an Architectural Masterpiece* (Florence: Mandragora, 2004).

31 Storia dell'arte, "Tempietto San Pietro in Montorio," https://www.progettostoriadellarte.it/2020/05/02/tempietto-san-pietro-in-montorio/ [accessed 2022].

32 N. H. Freeman, "Do children draw men with arms coming out of the head?," *Nature* 254 (1975), pp. 416–17; E. Burkitt et al., "The effect of affective characterizations on the size of children's drawings, *British Journal of Developmental Psychology* 21 (2003), pp. 565–84.

33 S. Toselli et al., "Growth of Chinese Italian infants in the first 2 years of life," *Annals of Human Biology* 32 (2005), pp. 15–29.

34 C. Alacevich and A. Tarozzi, "Child height and intergenerational transmission of health: Evidence from ethnic Indians in England," *Economics and Human Biology* 25 (2017), pp. 65–84.

35 Euromonitor International, *World Market for Luxury Goods* (London: Euromonitor International, 2022).

36 Madurodam, "Moet je meemaken," https://www.madurodam.nl [accessed 2022]; Fake Food Japan, "Small size replicas," https://fakefoodjapan.com/collections/small-size-replicas [accessed 2022].

37 Virtual tours are available at: https://www.louvre.fr/en/online-tours [accessed 2022] and https://www.museodelprado.es/en/whats-on/multimedia/visual-guide-to-the-prado-museum/4621ae59-3080-43bb-892b-34721f47ca96 [accessed 2021].

38 Guinness Book of World Records, "Most popular," https://www.guinnessworldrecords.com/records/showcase/most-popular [accessed 2022].

39 To see the most frightening images of tornadic storm clouds, search for "supercells" on the Web.

40 In 2020, *Ever ACE*—a 400-meter-long carrier with a capacity of 23,992 TEUs, built by Samsung Heavy Industries in South Korea—was the

world's largest container carrier: Marine Insight, "Top 10 world's largest container ships in 2022," June 11, 2021, https://www.marineinsight .com/know-more/top-10-worlds-largest-container-ships-in-2019/#1 _Ever_Ace.

41 The smallest discernible sizes depend on an individual's acuity as well as on available light, contrast, pattern, and color.

42 In extreme cases, tops of cumulonimbus clouds can be more than 20 kilometers above sea level.

43 FRED, "Real gross domestic product," https://fred.stlouisfed.org/ series/GDPC1; "Real gross domestic product per capita," https:// fred.stlouisfed.org/series/A939RX0Q048SBEA [accessed 2022].

44 D. W. Hone and M. J. Benton, "The evolution of large size: How does Cope's Rule work?," *Trends in Ecology and Evolution* 20 (2005), pp. 4–6.

45 N. A. Heim et al., "Cope's rule in the evolution of marine animals," *Science* 347 (2015), pp. 867–70.

46 F. A. Smith et al., "Body size evolution across the Geozoic," *Annual Review of Earth and Planetary Sciences* 44 (2016), pp. 523–53.

47 M. Monroe and F. Bokma, "Little evidence for Cope's Rule from Bayesian phylogenetic analysis of extant mammals," *Journal of Evolutionary Biology* 23 (2010), pp. 2017–21.

48 C. Pedrós-Alió and S. Manrubia, "The vast unknown microbial biosphere," *Proceedings of the National Academy of Sciences* 113 (2016), pp. 6585–7.

49 V. Smil, *Energy and Civilization: A History* (Cambridge, MA: MIT Press, 2017); V. Smil, *Making the Modern World: Materials and Dematerialization* (Chichester: John Wiley & Sons, 2013).

50 V. Smil, *Grand Transitions: How the Modern World Was Made* (New York: Oxford University Press, 2021).

51 In 1900, its annual increase was about 1.5 percent; at its peak, during the late 1960s, the rate was about 2.1 percent a year: V. Smil, *Growth: From Microorganisms to Megacities* (Cambridge, MA: MIT Press, 2019), p. 315.

52 Smil, *Prime Movers of Globalization.*

53 Power Technology, "Three Gorges Dam Hydro Electric Power Plant, China," https://www.power-technology.com/projects/gorges [accessed 2021].

54 V. Smil, *Still the Iron Age: Iron and Steel in the Modern World* (Amsterdam: Elsevier, 2016).

55 Burj Khalifa, "Facts & Figures," https://www.burjkhalifa.ae/en/the-tower/facts-figures/ [accessed 2022].

56 This size is for the Greater Tokyo Area or National Capital Region. Other definitions (taking into account functional or administrative boundaries) yield smaller population totals.

57 FRED. "Real gross domestic product," https://fred.stlouisfed.org/series/GDPC1 [accessed 2021].

58 Library of Congress, "The Library of Congress, 1800–1992," Jefferson's Legacy [website], https://www.loc.gov/loc/legacy/loc.html; M. Raymond, "How 'big' is the Library of Congress?," Library of Congress [blog], February 11, 2009, https://blogs.loc.gov/loc/2009/02/how-big-is-the-library-of-congress.

59 M. Lesk, "How much information is there in the world?" (1997), https://lesk.com/mlesk/ksg97/ksg.html.

60 V. Smil, *Transforming the Twentieth Century: Technical Innovations and Their Consequence* (New York: Oxford University Press, 2006).

61 N. Hoffmann, *Mergers and Acquisitions Strategy for Consolidations: Roll Up, Roll Out and Innovate for Superior Growth and Returns* (New York: McGraw Hill Education, 2012).

62 Institute of Mergers, Acquisitions and Alliances, "M&A Statistics," https://imaa-institute.org/mergers-and-acquisitions-statistics/ [accessed 2022].

63 Statista Research Department, "Global market share—statistics & facts," August 5, 2022, https://www.statista.com/topics/898/global-market-share/.

64 R. H. Casey, *The Model T: A Centennial History* (Baltimore, MD: Johns Hopkins University Press, 2008).

65 Autobytel, "Ten of the biggest SUVs," https://www.autobytel.com/luxury-sport-utility-vehicles/car-buying-guides/10-of-the-biggest-suvs-114176/ [accessed 2022].

66 The lightest Model T weighed 540 kilograms; the heaviest Yukon is 2,642 kilograms.

67 L. Cozzi and A. Petropoulos, "Growing preference for SUVs challenges emissions reductions in passenger car market," October 15, 2019, https://

www.iea.org/commentaries/growing-preference-for-suvs-challenges-emissions-reductions-in-passenger-car-market.

68 US Census Bureau, "Historical household tables," November 2021, https://www.census.gov/data/tables/time-series/demo/families/households.html.

69 R. Dietz, "New single-family home size trends lower," National Association of Home Builders, August 16, 2017, https://eyeonhousing.org/2017/08/new-single-family-home-size-trends-lower/.

70 These structures include many so-called McMansions, ostentatiously large houses of assorted pseudo-designs situated on very large lots.

71 But there are problems with these size measurements, with some specifications overstating the volume by nearly a third: D. Wroclawski, "Why refrigerator capacity claims don't add up," March 1, 2018, https://www.consumerreports.org/refrigerators/why-refrigerator-capacity-claims-dont-add-up/.

72 Sony Corporation, "Trinitron Color Video Monitor PVM-4300 Operating Manual," 1989, https://www.manualslib.com/manual/756743/Sony-Trinitron-Pvm-4300.html.

73 RTINGS.com., "TV size to distance calculator and science," March 12, 2021, https://www.rtings.com/tv/reviews/by-size/size-to-distance-relationship.

74 D. Moore, "Why aren't Americans happier than they were in the '70s?," Clearer Thinking, April 25, 2017, https://www.clearerthinking.org/post/2017/04/25/why-arent-americans-happier-than-they-were-in-the-70s.

75 M. Di Cesare et al., "The epidemiological burden of obesity in childhood: A worldwide epidemic requiring urgent action," *BMC Medicine* 17 (2019), p. 212.

76 US conscription lasted until June 1973. Men below 60 inches and above 80 inches (152.4 and 203.2 centimeters) were exempt. The Vietnam War (1964–1975) was the last conflict fought by conscripts; its US battle deaths reached 47,434, with 153,303 wounded: Department of Veterans Affairs, "America's Wars," 2020, https://www.va.gov/opa/publications/factsheets/fs_americas_wars.pdf.

77 For inequality estimates (Gini coefficient) by country see: World Bank, "Gini index (World Bank Estimate)," https://data.worldbank.org/indicator/SI.POV.GINI [accessed 2021].

78 As always, there are exceptions, as history makes some small countries (or regions) persistent locations of instability and conflicts whose impact reaches far beyond their limited areas: Israel, Lebanon, Kashmir, and Taiwan are obvious examples.

79 R. K. Kopparapu et al., "Habitable zones around main sequence stars: Dependence on planetary mass," *Astrophysical Journal Letters* 787 (2014), L29.

80 C. Ptolemaus, *Claudii Ptolemaei Opera quae Exstant Omnia* (Leipzig: Teubner, 1893, 1952).

81 A. van Helden, *Measuring the Universe: Cosmic Dimensions from Aristarchus to Halley* (Chicago: Chicago University Press, 1985).

82 M. Bucciantini et al., *Galileo's Telescope: A European Story* (Cambridge, MA: Harvard University Press, 2015).

83 J. Kepler, *Epitome Astronomiae Copernicanae* (Linz: Johannes Plancus, 1618).

84 D. W. Hughes, "Six stages in the history of the astronomical unit," *Journal of Astronomical History and Heritage* 4 (2001), pp. 15–28. The accurate distance was settled by the International Astronomic Union in 1964 and revised in 1976, with one astronomic unit equal to 23,454.78 Earth's radii, putting the Sun about 150 million kilometers away.

85 M. J. Reid and K. M. Menten, "The first stellar parallaxes revisited" (September 2020), https://arxiv.org/abs/2009.11913; J. D. Fernie, "The historical search for stellar parallax," *Journal of the Royal Astronomical Society of Canada* 69 (1975), pp. 153–61.

86 H. Shapley and H. D. Curtis, "The scale of the universe," *Bulletin of the National Research Council* 2/169 (1921), pp. 171–217; E. P. Hubble, "A spiral nebula as a stellar system, Messier 31," *The Astrophysical Journal* 69 (1929), p. 103.

87 D. N. Page, "Ab initio estimates of the size of the observable universe," *Journal of Cosmology and Astroparticle Physics* (2011).

88 B. Amos, "Lessons from the history of light microscopy," *Nature Cell Biology* 2 (2000), E151–2.

89 D. Bardell, "The invention of the microscope," *Bios* 75 (2004), pp. 78–84.

90 R. Hooke, *Micrographia* (London: Jo. Martyn and Ja. Allestry, Printers to the Royal Society, 1665).

91 I. Lawson, "Crafting the microworld: How Robert Hooke constructed knowledge about small things," *Notes and Records of the Royal Society* 70 (2016), pp. 23–44.

92 A. Leewenhoeck, "Observation, communicated to the publisher by Mr. Antony van Leewenhoeck, in a Dutch letter of the 9 October 1676 here English'd: concerning little animals by him observed in rain-well-sea and snow water; as also in water wherein pepper had lain infused," *Philosophical Transactions* 12 (1677), pp. 821–31; see also N. Lane, "The unseen world: Reflections on Leeuwenhoek (1677) 'Concerning little animals,'" *Philosophical Transactions of the Royal Society B* 370 (2015).

93 B. Ford, *The Leeuwenhoek Legacy* (Bristol: Biopress and Farand, 1991).

94 Cell Biology by the Numbers, "How big are viruses?," http://book.bionumbers.org/how-big-are-viruses/ [accessed 2022].

95 E. Ruska, "The development of the electron microscope and of electron microscopy," Nobel lecture, December 8, 1986, https://www.nobelprize.org/uploads/2018/06/ruska-lecture.pdf.

96 J. Kuo (ed.), *Electron Microscopy: Methods and Protocols* (Berlin: Springer, 2014).

97 R. E. Gordon, "Electron microscopy: A brief history and review of current clinical applications," *Methods in Molecular Biology* 1180 (2014), pp. 119–35.

II PERCEPTIONS, ILLUSIONS, MEASUREMENTS

1 M. Merleau-Ponty, *Phénoménologie de la perception* (Paris: Gallimard, 1945) [*Phenomenology of Perception*, trans. D. Landes (London: Routledge, 2012)]; L. R. Harris et al., "How our body influences our perception of the world," *Frontiers in Psychology* 6 (2015), pp. 819.

2 L. F. Jacobs et al., "Olfactory orientation and navigation in humans," *PLos ONE* 16/6 (June 2015).

3 Y. Tuan, *Space and Place: The Perspective of Experience* (Minneapolis: University of Minnesota Press, 1977), pp. 36, 44–5.

4 I. Gallagher, "Philosophical conceptions of the self: Implications for cognitive science," *Trends in Cognitive Science* 4 (2000), pp. 14–21.

5 J. Willis and A. Todorov, "First impressions: making up your mind after a 100-ms exposure to a face," *Psychological Science* 17 (2006), pp. 592–8.

6 F. F. Alsulaimani and W. Batwa, "Incisors' proportions in smile esthetics," *Journal of Orthodontic Science* 2 (2013), pp. 109–12.

7 Unfortunately, obesity in America has become both more common and more severe. Between 1988 and 1994 about 23 percent of all adults were obese and 2.8 percent severely obese (BMI > 40); by 2017–18 the respective shares were 42 percent and just over 9 percent: C. D. Fryar et al., "Prevalence of overweight, obesity, and severe obesity among adults aged 20 and over: United States, 1960–1962 through 2017–2018," NCHS Health E-Stats, January 29, 2021, https://www.cdc.gov/nchs/data/hestat/obesity-adult-17-18/obesity-adult.htm.

8 The primary reason for the difference is the structure of traditional Italian meals, with their sequence of relatively small servings of *antipasto*, *primo*, *secondo*, and *contorno*, rather than offering a small mountain of meaty pasta in a bowl, as many American restaurants would do.

9 During the late 1990s the average US hotel room was just over 30 square meters; new hotels have much smaller rooms, typically a bit less than 20 square meters but that is still well above the Japanese *bizunesu hoteru* where rooms are 13–15 square meters: Queviv, "Hotel rooms 20 years ago were twice as large as some of today's offerings," *USA Today*, November 4, 2015, https://www.usatoday.com/story/travel/roadwarriorvoices/2015/11/04/hotel-rooms-20-years-ago-were-twice-as-large-as-some-of-todays-offerings/83847338/.

10 C. Coosje van Bruggen, *Frank O. Gehry: Guggenheim Museum Bilbao* (New York: Solomon R Guggenheim Museum, 1997).

11 Guggenheim Bilbao, "Puppy: Jeff Koons," https://www.guggenheim-bilbao.eus/en/the-collection/works/puppy [accessed 2021].

12 Dalí's oozing watch paintings are among the most reproduced works of art. For Magritte see: X. Cannone, *René Magritte: The Revealing Image* (Brussels: Ludion, 2017).

13 S. E. Palmer et al., "Canonical perspective and the perception of objects," in J. Longand and A. Baddeley (eds), *Attention and Performance IX* (Hillsdale, NJ: Erlbaum, 1981), pp. 135–51.

14 D. I. Perrett et al., "Use of preferential inspection to define the viewing sphere and characteristic views of an arbitrary machined tool part," *Perception* 21 (1992), pp. 497–515.

15 V. Blanz et al., "What object attributes determine canonical views?," *Perception* 28 (1999), pp. 575–99.

16 E. Mezuman and Y. Weiss, "Learning about canonical views from internet image collections," *NIPS'12: Proceedings of the 25th International Conference on Neural Information Processing Systems* 1 (2012), pp. 1719–72.

17 T. Konkle and A. Oliva, "Canonical visual size for real-world objects," *Journal of Experimental Psychology: Human Perception and Performance* 37 (2011), pp. 23–37.

18 Musei Capitolini, "Colossal head of Constantine," http://capitolini .info/scu00757/?lang=en [accessed 2021].

19 Rhodes Guide, "The Colossus of Rhodes, a wonder of the ancient world," https://www.rhodesguide.com/travelguide/colossus_rhodes .php [accessed 2021].

20 Città di Firenze, "Palazzo Vecchio, Salone dei Cinquecento," https:// www.comune.fi.it/pagina/sale-monumentali/palazzo-vecchio [accessed 2021].

21 B. Chardère, *Les images des Lumière* (Paris: Gallimard, 1995).

22 E. Huhtamo, "Gulliver in figurine land," *Mediamatic* 4 (1990), pp. 101–5.

23 For a multidisciplinary perspective on screens, see: D. Chateau and J. Moure (eds), *Screens: From Materiality to Spectatorship—A Historical and Theoretical Reassessment* (Amsterdam: Amsterdam University Press, 2016).

24 E. Lampert-Greaux, "Obscura projects on the Empire State Building," LiveDesign, April 19, 2017, https://www.livedesignonline.com/ installations/obscura-projects-empire-state-building.

25 J. Verne, "In the year 2889," *Forum* 6 (1889), pp. 662–77; E. Huhtamo, "The sky is (not) the limit: Envisioning the ultimate public media display," *Journal of Visual Culture* 8 (2010), pp. 329–48.

26 By 2019, TVs with a 65-inch (165-centimeter) diagonal became the most popular US size: J. Porter, "65-inch TVs are now the most popular choice for North American households: report," The Verge, July 5, 2019, https://www.theverge.com/2019/7/5/20682913/most-popular-tv-size-65-inch-55-preference-market-research; see also D. Pogue, "A brief history of aspect ratios, aka screen proportions," *Scientific American*, February 20, 2018, https://www.scientificamerican.com/article/a-brief-history-of-aspect-ratios-aka-screen-proportions/.

27 J. Shi et al., "Understanding the lives of problem gamers: The meaning, purpose, and influences of video gaming," *Computers in Human Behavior* 97 (2019), pp. 291–303; M. Zastrow, "Is video game addiction really an addiction?," *Proceedings of the National Academy of Sciences* 114 (2017), pp. 4268–72.

28 G. Lissak, "Adverse physiological and psychological effects of screen time on children and adolescents: Literature review and case study," *Environmental Research* 164 (2018), pp. 149–57; T. J. Saunders et al., "Screen time and health indicators among children and youth: Current evidence, limitations and future directions," *Applied Health Economics and Health Policy* 15 (2017), pp. 323–31.

29 F. C. Müller-Lyer, "Optische Urteilstäuschunge," *Archiv für Physiologie Suppl.* (1889), pp. 263–70; H. Ebbinghaus, *Grundzüge der Psychologie* (Leipzig: Veit & Co., 1902). For a systematic review and typology of visual illusions, see D. Todorović, "What are visual illusions?," *Perception* 49 (2020), pp. 1128–99.

30 C.-C. Carbon, "The folded paper size illusion: Evidence of inability to perceptually integrate more than one geometrical dimension," *i-Perception* (July–August 2016), pp. 1–5.

31 T. Leibovich et al., "Itsy bitsy spider? Valence and self-relevance predict size estimation," *Biological Psychology* 121 (2016), pp. 138–45; M. W. Vasey, "It was as big as my head, I swear! Biased spider size estimation in spider phobia," *Journal of Anxiety Disorders* 26 (2012), pp. 20–4; Y. Shibana, "Treatment effect on biases in size estimation in spider phobia," *Biological Psychology* 121 (2016), pp. 146–52.

32 B. A. Teachman, "A new mode of fear expression: Perceptual bias in height fear," *Emotion* 8 (2008), pp. 296–301; J. K. Stefanucci and D. R. Proffitt, "The roles of altitude and fear in the perception of height," *Journal of Experimental Psychology: Human Perception and Performance* 35 (2009), pp. 424–38.

33 J. S. Bruner and C. C. Goodman, "Value and need as organizing factors in perception," *Journal of Abnormal and Social Psychology* 42 (1947), pp. 33–44; D. Dubois et al., "The accentuation bias: Money literally looms larger (and sometimes smaller) to the powerless," *Social Psychological and Personality Science* 1/3 (2010), pp. 199–205.

34 T. A. Salthouse et al., "An illusion of ingestion," *Perception & Psychophysics* 27 (1980), pp. 564–8.

35 R. Weidner et al., "The Moon Illusion and size-distance scaling: Evidence for shared neural patterns," *Journal of Cognitive Neuroscience* 26 (2014), pp. 1871–82.

36 L. R. Harris and C. Mander, "Perceived distance depends on the orientation of both the body and the visual environment," *Journal of Vision* 14 (2014), pp. 1–8.

37 G. A. Radvansky and L. A. Carlson-Radvansky, "Uncertainty in estimating distances from memory," *Memory & Cognition* 23 (1995), pp. 596–606.

38 R. Volcic et al., "Visuomotor adaptation changes stereoscopic depth perception and tactile discrimination," *The Journal of Neuroscience* 33 (2014), pp. 17081–8.

39 G. Clément et al., "Distance and size perception in astronauts during long-duration spaceflight," *Life* 3 (2013), pp. 524–37.

40 R. W. Baird and S. M. Burkhart, "Bias and variability in distance estimation on the water: Implications for the management of whale watching," IWC Meeting Document SC/52/WW1 (2000).

41 C. Button et al., "Distance perception in an open water environment: Analysis of individual differences," *Attention, Perception & Psychophysics* 78 (2016), pp. 915–22.

42 C. A. Meissner et al., "Person Descriptions as Eyewitness Evidence," in R. C. L. Lindsay et al. (eds), *The Handbook of Eyewitness Psychology, Vol. 2. Memory for People* (Mahwah, NJ: Lawrence Erlbaum Associates Publishers, 2007), pp. 3–34; T. D. Albright, "Why eyewitnesses fail," *Proceedings of the National Academy of Sciences* 114 (2017), pp. 7758–64.

43 R. H. Flin and J. W. Shepherd, "Tall stories: Eyewitnesses' ability to estimate height and weight characteristics," *Human Learning* 5 (1986), pp. 29–38.

44 R. C. L. Lindsay et al., "How variations in distance affect eyewitness reports and identification accuracy," *Law and Human Behavior* 32 (2008), pp. 526–35.

45 A. Tajadura-Jiménez et al., "Embodiment in a child-like talking virtual body influences object size perception, self-identification, and subsequent real speaking," *Scientific Reports* 7 (2017), article no: 9637. See also D. Banakou et al., "Illusory ownership of a virtual child body causes overestimation of object sizes and implicit attitude changes," *Proceedings*

of the National Academy of Sciences 110 (2013), pp. 12846–51; B. van der Hoort et al., "Being Barbie: The size of one's own body determines the perceived size of the world," *PLoS ONE* 6/5 (2011), 6:e20195.

46 S. A. Linkenauger et al., "Welcome to Wonderland: The influence of the size and shape of a virtual hand on the perceived size and shape of virtual objects," *PLoS ONE* 8/7 (2013), e68594; N. Ogawa et al., "Distortion in perceived size and body-based scaling in virtual environments," *8th Augmented Human International Conference* (2017).

47 K. Opichka and C. Smith, "Accuracy of self-reported heights and weights in a predominately low-income, diverse population living in the USA," *American Journal of Human Biology* 30/6 (2018), e23184.

48 C. O. Chigbu et al., "Impact of perceptions of body size on obesity and weight management behaviour: A large representative population study in an African setting," *Journal of Public Health* 43 (2019), e54–e61.

49 A. W. Y. Chan et al., "Misalignment between perceptual boundaries and weight categories reflects a new normal for body size perception," *Scientific Reports* 11 (2021), p. 10442.

50 J. Allen and G. C. Prkachin, "Parental awareness and perception of their children's body size," *Open Journal of Medical Psychology* 2 (2013), pp. 77–80.

51 M. J. Tovée et al., "Healthy body equals beautiful body? Changing perceptions of health and attractiveness with shifting socio-economic status," in V. Swami and A. Furnham (eds), *Body Beautiful: Evolutionary and Sociocultural Perspectives* (Basingstoke, UK: Palgrave Macmillan, 2007), pp. 108–28; S. Grabe et al., "The role of the media in body image concerns among women: A meta-analysis of experimental and correlational studies," *Psychological Bulletin* 134 (2008), pp. 460–76.

52 C. Winkler and G. Rhodes, "Perceptual adaptation affects attractiveness of female bodies," *British Journal of Psychology* 96 (2005), pp. 141–54.

53 S. K. Madsen et al., "Visual processing in anorexia nervosa and body dysmorphic disorder: Similarities, differences, and future research directions," *Journal of Psychiatric Research* 47 (2013), 1483e1491.

54 R. Zopf et al., "Body distortions in anorexia nervosa: Evidence for changed processing of multisensory bodily signals," *Psychiatry Research* 245 (2016), pp. 473–81; H. W. Hoek, "Incidence, prevalence and

mortality of anorexia nervosa and other eating disorders," *Current Opinions in Psychiatry* 19 (2006), pp. 389–94.

55 T. Brockmeyer et al., "Advances in the treatment of anorexia nervosa: A review of established and emerging interventions," *Psychological Medicine* 48 (2018), pp. 1228–56.

56 A. S. Bjornsson, "Body dysmorphic disorder," *Dialogues in Clinical Neuroscience* 12 (2010), pp. 221–32.

57 N. A. Vashi, "Obsession with perfection: Body dysmorphia," *Clinics in Dermatology* 34 (2016), pp. 788–91.

58 G. M. van Koningsbruggen, "Through the eyes of dieters: Biased size perception of food following tempting food primes," *Journal of Experimental Social Psychology* 47 (2011), pp. 293–9.

59 M. Condrasky, "Chefs' opinions of restaurant portion sizes," *Obesity* 15 (2007), pp. 2086–94.

60 R. Klara, "Table the issue," *Restaurant Business* 103 (2004), pp. 14–15.

61 B. Wansink and K. van Ittersum, "Portion size me: Plate-size induced consumption norms and win-win solutions for reducing food intake," *Journal Exp Psychol App* 19/4 (2013), pp. 320–32. Earlier, the same authors found a similar effect with glasses: children and adults given a short wide glass poured themselves more juice than those given a tall, slender glass—but perceived the opposite to be true! "Bottoms up! The influence of elongation on pouring and consumption volume," *Journal of Consumer Research* 30 (2003), pp. 450–63.

62 S. Nicolas, "Joseph Delboeuf on visual illusions: A historical sketch," *The American Journal of Psychology* 108 (1995), pp. 563–74.

63 E. Libotte et al., "The influence of plate size on meal composition: Literature review and experiment," *Appetite* 82 (2014), pp. 91–6.

64 J. H. Williams, *Defining and Measuring Nature* (San Rafael, CA: Morgan & Claypool Publishers, 2014).

65 K. Spence, "Ancient Egyptian chronology and the astronomical orientation of pyramids," *Nature* 408 (2000), pp. 321–4.

66 E. Hadingham, "Unlocking mysteries of the Parthenon," *Smithsonian Magazine* (February 2008), https://www.smithsonianmag.com/history/unlocking-mysteries-of-the-parthenon-16621015/.

67 When Roman power was at its peak, the empire controlled about 15 percent of what it knew to be the extent of the inhabited Old World,

and just 3 percent of the entire continental surface: V. Smil, *Why America Is Not a New Rome* (Cambridge, MA: MIT Press, 2010).

68 There are only two other countries that are not officially metric: Liberia and Myanmar. Not surprisingly, the generations-long resistance to the metric system is now portrayed as yet another demonstration of America's love of freedom.

69 A. T. Steegmann, "18th century British military stature: Growth cessation, selective recruiting, secular trends, nutrition at birth, cold and occupation," *Human Biology* 57 (1985), pp. 77–95.

70 G.-L. L. de Buffon, *Histoire naturelle: Supplement: Tome quatrieme* (Paris: Imprimerie Royale, 1753).

71 L. R. Villermé, "Mémoire sur la taille de l'homme en France," *Annales d'HygiènePublique et de Médicine Légale* 1 (1829; K. Staub et al., "Edouard Mallet's early and almost forgotten study of the average height of Genevan conscripts in 1835," *Economics & Human Biology* 9 (2011), pp. 438–42; A. Quetelet, *Sur l'homme et le développement de ses facultés*, vol. 2 (Paris: Bachelier, 1835).

72 H. P. Bowditch, "The Growth of Children Studied by Galton's Percentile Grades," in *22nd Annual Report of the State Board of Health of Massachusetts* (Boston: Wright & Potter, 1891), pp. 479–525; J. M. A. Tanner, "Concise history of growth studies from Buffon to Boas," in F. Falkner and J. M. A. Tanner (eds), *Human Growth, Volume 3, Neurobiology and Nutrition* (Berlin: Springer Verlag, 1979), pp. 515–93.

73 WHO, *Child Growth Standards* (Geneva: WHO, 2006); CDC "Growth charts," 2010, https://www.cdc.gov/growthcharts/.

74 T. Cuff, "Historical Anthropometrics," EH.net, https://eh.net/encyclopedia/historical-anthropometrics/; J. Komlos, "Anthropometric history: An overview of a quarter century of research," *Anthropologischer Anzeiger* 67 (2009), pp. 341–56.

75 R. Floud et al., *The Changing Body: Health, Nutrition, and Human Development in the Western World since 1700* (Cambridge: Cambridge University Press, 2011); R. W. Fogel, *Explaining Long-Term Trends in Health and Longevity* (Cambridge: Cambridge University Press, 2012).

76 R. H. Steckel, "New light on the 'Dark Ages': The remarkably tall stature of northern European men during the medieval era," *Social Science History* 28 (2004), pp. 211–29.

77 N. Koepke and J. Baten, "The biological standard of living in Europe during the last two millennia," *European Review of Economic History* 9 (2005), pp. 61–95; G. Kron, "Anthropometry, physical anthropology, and the reconstruction of ancient health, nutrition, and living standards," *Historia* 54 (2005), pp. 68–83.

78 T. J. Hatton and B. E. Bray, "Long run trends in the heights of European men, 19th–20th centuries," *Economics & Human Biology* 8 (2010), pp. 405–13.

79 H. de Beer, "Observations on the history of Dutch physical stature from the late-Middle Ages to the present," *Economics and Human Biology* 2 (2003), pp. 45–55.

80 NCD Risk Factor Collaboration (NCD-RisC), "A century of trends in adult human height," *eLife* 5 (2016), e13410.

81 M. Hermanussen et al., "Height and skeletal morphology in relation to modern life style," *Journal of Physiological Anthropology* 34 (2015), p. 41.

82 C. Ji and T. Chen, "Secular changes in stature and body mass index for Chinese youth in sixteen major cities, 1950s–2005," *American Journal of Human Biology* 20 (2008), pp. 530–7.

83 World Population Review, "Average height by country," https://worldpopulationreview.com/country-rankings/average-height-by-country [accessed 2021]; P. Grasgruber et al., "The coast of giants: An anthropometric survey of high schoolers on the Adriatic coast of Croatia," *PeerJ* 7 (2019), e6598.

84 R. H. Steckel, "Heights and human welfare: Recent developments and new directions," *Explorations in Economic History* 46 (2009), pp. 1–23.

85 P. M. Visscher et al., "From Galton to GWAS: Quantitative genetics of human height," *Genetic Research* 92 (2010), pp. 371–9; E. A. Boyle et al., "An expanded view of complex traits: From polygenic to omnigenic," *Cell* 169 (2017), pp. 1177–86.

86 R. H. Steckel, "Heights and human welfare: Recent developments and new directions," *Explorations in Economic History* 46 (2009), pp. 1–23; A. Singh-Manoux, "Trends in the association between height and socioeconomic indicators in France, 1970–2003," *Economics and Human Biology* 8 (2010), pp. 396–404; R. W. Fogel, *Explaining Long-Term Trends in Health and Longevity* (Cambridge: Cambridge University Press, 2012).

87 P. K. Bird et al., "Income inequality and social gradients in children's height: A comparison of cohort studies from five high-income countries," *British Medical Journal Paediatrics Open* 3 (2019), e000568.

88 A. Case and C. Paxson, "Stature and status: Height, ability, and labor market outcomes," *Journal of Political Economy* 116 (2008), pp. 499–532.

89 S. Reynolds et al., "Disparities in children's vocabulary and height in relation to household wealth and parental schooling: A longitudinal study in four low- and middle-income countries," *SSM—Population Health* 3 (2017), pp. 767–86.

90 F. Cinnirella et al., "Why does height matter for educational attainment? Evidence from German children," *Economics and Human Biology* 9 (2011), pp. 407–18.

91 J. M. Sundet et al., "Resolving the genetic and environmental sources of the correlation between height and intelligence: A study of nearly 2,600 Norwegian male twin pairs," *Twin Research and Human Genetics* 8 (2005), pp. 307–11; M. C. Keller et al., "The genetic correlation between height and IQ: Shared genes or assortative mating?," *PLoS Genetics* 9/4 (2013), e1003451.

92 E. B. Gowin, *The Executive and His Control of Men* (New York: Macmillan, 1915).

93 A. Croppenstedt and C. Muller, "The impact of farmers' health and nutritional status on their productivity and efficiency: Evidence from Ethiopia," *Economic Development and Cultural Change* 48 (2000), pp. 475–502; S. Dinda et al., "Height, weight and earnings among coalminers in India," *Economics and Human Biology* 4 (2006), pp. 342–50; G. Heineck, "Up in the skies?: The relationship between body height and earnings in Germany," *Labour* 19 (2005), pp. 469–89; M. Kortt and A. Leigh, "Does size matter in Australia?," *Economic Record* 86 (2010), pp. 71–83; K. Sohn, "The value of male height in the marriage market," *Economics and Human Biology* 18 (2015), pp. 110–24; W. Gao and R. Smyth, "Health human capital, height and wages in China," *Journal of Development Studies* 46 (2009), pp. 466–84; E. Yamamura et al., "Decomposing the effect of height on income in China: The role of market and political channels," *Economics and Human Biology* 19 (2015), pp. 62–74; D. LaFavea and D. Thomas, "Height and cognition at work: Labor market productivity in a low income setting," *Economics and Human Biology* 25 (2017), pp. 52–64.

94 A. Schick and R. H. Steckel, "Height, human capital, and earnings: The contributions of cognitive and noncognitive ability," *Journal of Human Capital* 9 (2015), pp. 94–115.

95 C. Auld, "Global country-level estimates of associations between adult height and the distribution of income," *American Journal of Human Biology* 30 (2018), e23152.

96 R. Adams et al., *Are CEOs Born Leaders?: Lessons from Traits of a Million Individuals* (Helsinki: Aalto School of Business, 2016).

97 E. Yamamura et al., "Decomposing the effect of height on income in China: The role of market and political channels," *Economics and Human Biology* 19 (2015), pp. 62–74.

98 J. Wang et al., "What is creating the height premium? New evidence from a Mendelian randomization analysis in China," *PLoS ONE* 15/4 (2020), e0230555.

99 In 2016, the NCD-RisC report opened with this statement: "Being taller is associated with enhanced longevity, lower risk of adverse pregnancy outcomes and cardiovascular and respiratory diseases, and higher risk of some cancers." NCD Risk Factor Collaboration (NCD-RisC, "A century of trends in adult human height," *eLife* 5 (2016), e13410. The first claim must be reversed.

100 T. T. Samaras and L. H. Storm, "Impact of height and weight on life span," *Bulletin of the World Health Organization* 70 (1992); T. T. Samaras, "How height is related to our health and longevity: A review," *Nutrition and Health* 21 (2012), pp. 247–61.

101 S. Lemez et al., "Do 'big guys' really die younger? An examination of height and lifespan in former professional basketball players," *PLoS ONE* 12/10 (2017), e0185617.

102 J. Green et al., "Height and cancer incidence in the Million Women Study: Prospective cohort, and meta-analysis of prospective studies of height and total cancer risk," *Lancet Oncology* 12 (2011), pp. 785–94; E. Benyi et al., "Adult height is associated with risk of cancer and mortality in 5.5 million Swedish women and men," *Journal of Epidemiology & Community Health* 73 (2019), pp. 730–6; Y. J. Choi et al., "Adult height in relation to risk of cancer in a cohort of 22,809,722 Korean adults," *British Journal of Cancer* 120 (2019), pp. 668–74; E. Giovanucci, "A growing link—what is the role of height in cancer risk?," *British Journal of Cancer* 120 (2019), pp. 575–6.

103 L. Nunney, "Size matters: Height, cell number and a person's risk of cancer," *Proceedings of the Royal Society B* 285 (2018), article 20181743.

104 J. Ayuk and M. C. Sheppard, "Does acromegaly enhance mortality?," *Review of Endocrinology and Metabolic Disorders* 9 (2008), pp. 33–9.

105 D. I. Silverman et al., "Life expectancy in the Marfan syndrome," *American Journal of Cardiology* 15 (1995), article 1571060.

III PROPORTIONS, SYMMETRY, AND ASYMMETRY

1 *National Geographic* posted a video showing a 270-kilogram octopus shapeshift itself to wriggle through a passageway the size of a US quarter—that is, the diameter of a mere 24.26 millimeters! National Geographic, "Octopus Escape" [video], https://www.youtube.com/watch?v=SCAIedFgdYo.

2 K. Cracknell et al., "Pentaradial eukaryote suggests expansion of suspension feeding in White Sea-aged Ediacaran communities," *Scientific Reports Sci Rep* 11 (2021), article 4121.

3 The tallest cruise ships that sail through the canal are nearly 67 meters tall, compared to 43 meters for the cupola of San Marco and 98.6 meters for the Campanile. Few images of contrasting sizes equal the view of these giant ships squeezed inside the canal, against the background of the sinking city.

4 For an extensive and well-illustrated commentary on Dürer's work see: *Gallucci's Commentary on Dürer's Four Books on Human Proportion*, trans. James Hutson and available at https://www.openbookpublishers.com/books/10.11647/obp.0198.

5 Merriam-Webster, "Proportion," https://www.merriam-webster.com/dictionary/proportion.

6 W. L. Jungers et al., "The evolution of body size and shape in the human career," *Philosophical Transactions of the Royal Society B* 371 (2016), article 20150247.

7 W. Wang and R. H. Crompton, "The role of load-carrying in the evolution of modern body proportions," *Journal of Anatomy* 204 (2004), pp. 417–30.

8 W. L. Jungers et al., "The evolution of body size and shape in the human career," *Philosophical Transactions of the Royal Society B* 371 (2016), article 20150247.

9 M. Will et al., "Long-term patterns of body mass and stature evolution within the hominin lineage," *Royal Society Open Science* 4 (2017), article 171339.

10 C. Bergmann, "Ueber die Verhältnisse der Wärmoekonomie der Thiere zu ihrer Grösse," *Göttinger Studien* 3 (1847), pp. 595–708; J. A. Allen, "The influence of physical conditions in the genesis of species," *Radical Review* 1 (1877), pp. 108–40.

11 M. J. Tilkens et al., "The effects of body proportions on thermoregulation: An experimental assessment of Allen's rule," *Journal of Human Evolution* 53 (2007), article 286e291.

12 The most obvious examples are southern Sudan's Dinka and Kenya's Maasai, both also being beneficiaries of high-protein, cattle-based (milk, blood, meat) diets.

13 E. Pomeroy et al., "Population history and ecology, in addition to climate, influence human stature and body proportions," *Scientific Reports* 11 (2021), p. 274.

14 M. A. Little, "Evolutionary strategies for body size," *Frontiers of Endocrinology* 11 (2020), p. 107.

15 M. Yokoya and Y. Higuchi, "Day length may make geographical difference in body size and proportions: An ecological analysis of Japanese children and adolescents," *PLoS ONE* 14/1 (2019), e0210265.

16 By far the most detailed attempt to reconstruct the lost canon is R. Tobin, "The Canon of Polykleitos," *American Journal of Archaeology* 79 (1975), pp. 307–21.

17 LeonardodaVinci.Net, "The Vitruvian Man—by Leonardo da Vinci," https://www.leonardodavinci.net/the-vitruvian-man.jsp.

18 Vitruvius Pollio, Chapter I: "On Symmetry: In Temples and In the Human Body," *The Ten Books on Architecture*, ed. M. H. Morgan, Book III, http://www.perseus.tufts.edu/hopper/text?doc=Vitr.%20 3.1&lang=original.

19 P. Sorokowski and B. Pawłowski, "Adaptive preferences for leg length in a potential partner," *Evolution and Human Behavior* 29/2 (2007).

20 J. C. K. Wells et al., "Associations of stunting at 2 years with body composition and blood pressure at 8 years of age: Longitudinal cohort analysis from lowland Nepal," *European Journal of Clinical Nutrition* 73 (2019), pp. 302–10.

21 P. Sorokowski et al., "Attractiveness of leg length: Report from 27 nations," *Journal of Cross-Cultural Psychology* 42 (2011), pp. 131–9.

22 T. M. M. Versluys et al., "The influence of leg-to-body ratio, arm-to-body ratio and intra-limb ratio on male human attractiveness," *Royal Society Open Science* 5 (2018), article 171790; S. Kiire, "Effect of leg-to-body ratio on body shape attractiveness," *Archives of Sex Behavior* 45 (2016), pp. 901–10.

23 V. Svami, "The leg-to-body ratio as a human aesthetic criterion," *Body Image* 3 (2006), pp. 317–23; V. Svami et al., "Cultural significance of leg-to-body ratio preferences? Evidence from Britain and rural Malaysia," *Asian Journal of Social Psychology* 10 (2007), pp. 265–9.

24 B. Bogin and M. I. Varela-Silva, "Leg length, body proportion, and health: A review with a note on beauty," *International Journal of Environmental Research and Public Health* 7 (2010), pp. 1047–75.

25 M. H. McIntyre, "Adult stature, body proportions and age at menarche in the United States National Health and Nutrition Examination Survey (NHANES) III," *Annals of Human Biology* 38 (2011), pp. 716–20.

26 L. Welters and A. Lillethun, *Fashion History: A Global View* (London: Bloomsbury Academic, 2018).

27 World Obesity, "Global Obesity Observatory," https://data.worldobesity.org [accessed 2021].

28 WHO, *Waist Circumference and Waist-hip Ratio: Report of a WHO Expert Consultation* (2008), meeting report, 16 May 2011, https://www.who.int/publications/i/item/9789241501491.

29 P. Srikanthan et al., "Waist-hip-ratio as a predictor of all-cause mortality in high-functioning older adults," *Annals of Epidemiology* 19 (2009), pp. 724–31.

30 A. Lehmann et al., "Temporal trends, regional variation and socio-economic differences in height, BMI and body proportions among German conscripts, 1956–2010," *Public Health Nutrition* 20 (2016), pp. 391–403.

31 G. Heineck, "Height and weight in Germany, evidence from the German Socio-Economic Panel," *Economics and Human Biology* 4 (2006), pp. 359–82.

32 J. T. Manning et al., "The ratio of 2nd to 4th digit length: A predictor of sperm numbers and concentrations of testosterone, luteinizing hormone and oestrogen," *Human Reproduction* 13 (1998), pp. 3000–4.

33 National Library of Medicine, "2D:4D," PubMed.gov, https://pubmed .ncbi.nlm.nih.gov/?term=2D%3A4D [accessed 2021].

34 Here are just three recent examples—looking at the ratio's link with schizophrenia, cancer, and Olympic performance: Y. Han et al., "Association between the 2D:4D ratio and schizophrenia," *Journal of Internal Medicine* 48 (2020), article 300060520929148; A. Bunevicius, "The association of digit ratio (2D:4D) with cancer: A systematic review and meta-analysis," *Disease Markers* (2018), article7698193; E. Eklund et al., "Digit ratio (2D:4D) and physical performance in female Olympic athletes," *Frontiers of Endocrinology* (May 2020).

35 L. Mitch, "Talk to the hand. Scientists try to debunk idea that finger length can reveal personality and health," *Science* (June 6, 2019), https:// www.science.org/content/article/talk-hand-scientists-try-debunk-idea-finger-length-can-reveal-personality-and-health.

36 L. Kratochvíl and J. Flegr, "Differences in the 2nd to 4th digit length ratio in humans reflect shifts along the common allometric line," *Biology Letters* 5/5 (2009).

37 S. E. Huber et al., "Prenatal androgen-receptor activity has organizational morphological effects in mice," *PLoS ONE* 12/11 (2017), e0188752.

38 Not surprisingly, association with COVID-19 became yet another 2D:4D topic—with no links found: J. A. L. Jaeger et al., "No credible evidence for links between 2D:4D and COVID-19 outcomes: A probabilistic perspective on digit ratio, ACE variants, and national case fatalities," *Early Human Development* 152 (2021), article 105272.

39 J. de Vries, "Luxury in the Dutch Golden Age in Theory and Practice," in M. Berg and E. Eger (eds), *Luxury in the Eighteenth Century* (London: Palgrave Macmillan, 2003), pp. 41–56.

40 N. Ireson and S. Fraquelli (eds), *Modigliani* (New York: Rizzoli, 2017).

41 A. Giusti, *Inganni ad arte: Meraviglie del trompe-l'oeil dall'antichità al contemporaneo* (Florence: Mandragora, 2009).

42 K. K. Butler et al., *Georges Braque and the Cubist Still Life, 1928–1945* (Munich: Prestel, 2013).

43 Emile Cardon, in *La Presse*, voiced the majority opinion about the first exhibition of impressionist paintings: "This school does away with two things: line, without which it is impossible to reproduce any form, animate or inanimate, and colour, which gives the form the appearance of reality . . . In examining the works exhibited . . . one wonders whether one is seeing the fruit either of a process of mystification which is highly unsuitable for the public, or the result of mental derangement which one could not but regret." See E. Cardon, "The exhibition of the Revoltes," *La Presse* (April 29, 1874), http://www .artchive.com/galleries/1874/74critic.htm#chesneau.

44 V. Goffaux, "Fixed or flexible? Orientation preference in identity and gaze processing in humans," *PLoS ONE* 14/1 (2019), e0210503.

45 X. Deng et al., "A 'wide' variety: Effects of horizontal versus vertical display on assortment processing, perceived variety, and choice," *Journal of Marketing Research* 53/5 (2016), pp. 682–98.

46 B. Lee et al., "Dissecting landscape art history with information theory," *Proceedings of the National Academy of Sciences* 117 (2020), pp. 26580–90.

47 G. Mather, "Aesthetic judgement of orientation in modern art," *Perception* 3 (2012), pp. 18–24.

48 You can see thousands of paintings at WikiArt: Visual Art Encyclopedia (https://www.wikiart.org) or at the Web Gallery of Art (https:// www.wga.hu/). If you prefer a single-volume book, this is the most convenient (albeit highly arguable) selection: S. Farthing, *1001 Paintings You Must See Before You Die* (New York: Universe Publishing, 2011).

49 In the Museo Reina Sofia in Madrid, Picasso's *Guernica* (7.77 × 3.49 meters) is among the most horizontal of big canvases, with a 2.22 ratio. In the Prado Bosch's *Garden of Earthly Delights* (a triptych of 3.89 × 2.2 meters) has a ratio of 1.76. In the Uffizi, Botticelli's *Birth of Venus* (1.72 × 2.78 meters) has the ratio of 1.61; and in Chicago (The Art Institute), Seurat's *Afternoon on Grand Jatte* (3.08 × 2.08 meters) is 1.48.

50 Vitruvius Pollio, "On Symmetry."

51 V. Trimble, "Astrophysical symmetries," *Proceedings of the National Academy of Sciences USA* 93 (1996), pp. 14221–4; T. M. Dame and P. Thaddeus, "A molecular spiral arm in the far outer galaxy," *Astrophysical Journals Letters* 734 (2011), L24.

52 M. E. Kellman, "Symmetry in chemistry from the hydrogen atom to proteins," *Proceedings of the National Academy of Sciences USA* 93 (1996), pp. 14287–94; P. A. van der Hel, "The influence of perception on the distribution of multiple symmetries in nature and art," *Symmetry* 3 (2011), pp. 54–71.

53 There is no finer illustrated guide to radiolarians than Ernst Haeckel's classic: *Die Radiolarien (Rhizopoda radiaria)* (Berlin: Georg Reimer, 1862). A complete download of his exquisite hand-drawn illustrations is available at: https://www.biodiversitylibrary.org/item/40590#page/4/mode/1up.

54 For Islamic ornaments, see D. Clevenot, *Ornament and Decoration in Islamic Architecture* (London: Thames Hudson, 2017). For Chinese lattices: D. S. Dye, *A Grammar of Chinese Lattice* (Cambridge, MA: Harvard University Press, 1937).

55 Hermann Weyl's excellent brief treatment remains my favorite book: *Symmetry* (Princeton, NJ: Princeton University Press, 1952). Publications that belong to the infatuated category include H. Genz, *Symmetrie, Bauplan der Natur* (Munich: Piper, 1987) and A. V. Voloshinov, "Symmetry as a superprinciple of science and art," *Leonardo* 29 (1996), pp. 109–13. Other notable symmetry books include: R. Joseph, *Symmetry in Science* (Berlin: Springer, 1995); I. Hargittai and M. Hargittai, *Symmetry: A Unifying Concept* (New York: Random House, 1996); M. du Sautoy, *Symmetry* (New York: Harper, 2008); M. Hargittai and I. Hargittai, *Visual Symmetry* (Singapore: World Scientific Publishing, 2009).

56 I. Hargittai, "Response to Peter van Sommers," *Leonardo* 29 (1996), p. 149.

57 N. van Melick et al., "How to determine leg dominance: The agreement between self-reported and observed performance in healthy adults," *PLoS ONE* 12/12 (2017), e0189876.

58 M. Y. Mommaerts and B. A. M. M. L. Moerenhout, "Ideal proportions in full face front view, contemporary versus antique," *Journal of Cranio-Maxillo-Facial Surgery* 39 (2011), 107e110.

59 A. Iglesias-Linares et al., "Common standards in facial esthetics: Craniofacial analysis of most attractive black and white subjects according to *People* Magazine during previous 10 years," *Journal of Oral and Maxillofacial Surgery* 69 (2011), e216-e224.

60 G. Rhodes et al., "Facial symmetry and the perception of beauty," *Psychonomic Bulletin & Review* 5 (1998), pp. 659–69; D. I. Perrett et al., "Symmetry and human facial attractiveness," *Evolution and Human Behavior* 20 (1999), pp. 295–307; B. Fink et al., "Facial symmetry and judgements of attractiveness, health and personality," *Personality and Individual Differences* 41 (2006), pp. 491–9; L. W. Simmons et al., "Are human preferences for facial symmetry focused on signals of developmental instability?," *Behavioral Ecology* 15 (2004), pp. 864–71.

61 Z. Lewandowski and A. Pisula-Lewandowska, "The influence of change in the size of face elements on the perception of a woman's portrait," *Homo* 59 (2008), pp. 253–60.

62 M. Ibáñez-Berganza et al., "Subjectivity and complexity of facial attractiveness," *Scientific Reports* 9 (2019), article 8364.

63 T. Valentine et al., "Why are average faces attractive? The effect of view and averageness on the attractiveness of female faces," *Psychonomic Bulletin & Review* 11 (2004), pp. 482–7.

64 D. W. Zaidel and M. Hessamian, "Asymmetry and symmetry in the beauty of human faces," *Symmetry* 2 (2010), pp. 136–49.

65 D. Hodgson, "The first appearance of symmetry in the human lineage: Where perception meets art," *Symmetry* 3 (2011), pp. 37–53.

66 By far the best modern, well-illustrated, single-volume history of architecture to be used for the study of design symmetry is F. D. K. Ching et al., *A Global History of Architecture* (Hoboken, NJ: John Wiley & Sons, 2011). Perhaps the best short paperback classic (still in print) dealing with proportions is N. C. Curtis, *The Secrets of Architectural Composition* (Cleveland, OH: J. J. Jansen, 1923).

67 A. Thalal et al., "Symmetry in art and architecture of the Western Islamic world," *Crystallography Reviews* 24 (2018), pp. 102–30.

68 U. Schober, *Castles and Palaces of Europe* (Lisse: Rebo International, 2005).

69 L. Trevisan, *Andrea Palladio: The Villas* (Leguzzano: Sassi Editore, 2008); J. Morrissey, *The Genius in the Design: Bernini, Borromini, and the Rivalry That Transformed Rome* (New York: William Morrow, 2005).

70 R. Zerbst, *Gaudí: The Complete Buildings* (Cologne: Taschen, 2005); A. von Vegesack, *Czech Cubism: Architecture, Furniture, and Decorative Arts, 1910–1925* (Princeton, NJ: Princeton Architectural Press, 1992); A. Tilch, *Bauhaus Architecture: 1919–1933* (Munich: Prestel, 2018); W. A. Storer, *The Architecture of Frank Lloyd Wright* (Chicago: Chicago University Press, 2017).

71 P. Goldberger, *Building Art: The Life and Work of Frank Gehry* (New York: Knopf, 2015); Zaha Hadid Architects, *Zaha Hadid Architects: Design as Second Nature* (Mexico City: RM/MUAC, 2019).

72 L. Marsili et al., "Unraveling the asymmetry of Mona Lisa smile," *Cortex* 120 (2019), pp. 607–10.

73 A. Vcherashniaja, "Dürer: Evolution of artistic self in 13 self-portraits," Arthive, February 11, 2021, https://arthive.com/publications/2426~Drer _evolution_of_artistic_self in_13_selfportraits.

74 P. White, "Differences over time in head orientation in European portrait paintings," *Laterality* 24 (2019), pp. 525–37.

75 Euclid, *Elements*, Book VI, definition 3, https://mathcs.clarku.edu/~ djoyce/java/elements/bookVI/defVI3.html.

76 C. Budd, "Myths of math: The Golden Ratio," Plus [website], February 23, 2020, https://plus.maths.org/content/myths-maths-golden-ratio.

77 By far the most comprehensive and rigorous mathematical treatment of the ratio in all of its guises is: A. S. Posamenter and I. Lehmann, *The Glorious Golden Ratio* (Amherst, NY: Prometheus Books, 2012).

78 G. Harary and A. Tal, "The natural 3D spiral," *Eurographics* 30 (2011); U. Mukhopadhyay, "Logarithmic spiral—a splendid curve," *Resonance* 9 (2010), pp. 39–45.

79 D. W. Thompson, *On Growth and Form* (Cambridge: Cambridge University Press, 1942), p. 757.

80 Precise size is 85.6×53.98 millimeters.

81 V. Gendelman, "How to use the Golden Ratio to create gorgeous graphic designs," Company Folders [blog], September 15, 2015, https://www .companyfolders.com/blog/golden-ratio-design-examples; J. Brownlee, "The Golden Ratio: Design's biggest myth," *Fast Company*, April 13, 2015, https://www.fastcompany.com/3044877/the-golden-ratio-designs-biggest-myth.

82 But only one among the world's most admired masterpieces (listed in the section on proportions), Botticelli's *Birth of Venus*, is painted on a canvas whose size is that of a golden rectangle.

83 G. Markowsky, "Misconceptions about the Golden Ratio," *The College Mathematics Journal* 23 (1992), pp. 2–19.

84 K. Devlin, "The myth that will not go away," Devlin's Angle [website], May 2007, https://www.maa.org/external_archive/devlin/devlin_05_07.html.

85 L. Pacioli, *Divina proportione* (Venice: A. Paganius Paganinus, 1509), available at: https://archive.org/details/divinaproportionoopaci.

86 M. Ohm, *Die reine Elementar-Mathematik* (Berlin: Jonas Verlag, 1815).

87 A. Zeising, *Neue Lehre von den Proportionen des menschlichen Körpers* (Leipzig: Rudolph Weigel, 1854).

88 G. T. Fechner, "Ueber die Frage des goldenen Schnittes," *Archiv für die zeichnende Künste* 11 (1865), pp. 100–12.

89 Germans continue to be the leading proponents of what they still see as the "mysterious" and universal key to everything: O. Götze and L. Kugler (eds), *Divine Golden Ingenious: The Golden Ratio as a Theory of Everything?* (Munich: Hirmer Publishers, 2016).

90 G. Yetkin et al., "Golden Ratio is beating in our heart," *International Journal of Cardiology* 168 (2020), pp. 4926–7.

91 Here are just three references to surprising claims, concerning reinforced concrete, sperm, and special relativity: A. P. Fantilli and B. Chiaia, "Golden ratio in the crack pattern of reinforced concrete structure," *Journal of Engineering Mechanics* 139 (2010), pp. 1178–84; W. E. Roudebush et al., "The golden ratio and an aesthetically pleasing sperm: Towards an objective assessment of sperm head morphology," *Fertility and Sterility* 86 (2006), pp. 59–68; L. di G. Sigalotti and A. Mejias, "The golden ratio and special relativity," *Chaso, Solitons and Fractals* 30 (2006), pp. 521–4.

92 C. D. Green, "All that glitters: A review of psychological research on the aesthetics of the g section," *Perception* 24 (1995), pp. 937–68.

93 M. K. Alam et al., "Multiracial facial golden ratio and evaluation of facial appearance," *PLoS ONE* 10/11 (2015), e0142914.

94 S. Polat et al., "The face shape and golden ratio classification in Turkish healthy population," *Journal of Evolution of Medical and Dental Sciences*

9/2 (2020), pp. 111–15; K. S. Jang et al., "A three-dimensional photo-grammetric analysis of the facial esthetics of the Miss Korea pageant contestants," *The Korean Journal of Orthodontics* 47 (2017), pp. 87–99.

95 C. Burusapat and P. Lekdaeng, "What is the most beautiful facial pro-portion in the 21st century? Comparative study among Miss Universe, Miss Universe Thailand, neoclassical canons, and facial golden ratios," *Plastic and Reconstructive Surgery—Global Open* 7/? (2019), c2044.

96 R. Fischler, "On the application of the golden ratio in the visual arts," *Leonardo* 14 (1981), pp. 31–2; R. Herz-Fischler, "An examination of claims concerning Seurat and 'The Golden Number,'" *Gazette des Beaux Arts* 125 (1983), pp. 109–12.

97 Government of Dubai, "Dubai Frame," https://www.dubaiframe.ae/en/about-us [accessed 2021]; Visit Dubai Frame, https://www.visit dubaiframe.com/.

IV SIZE DESIGNS:
THE GOOD, THE BAD, THE OUTRAGEOUS

1 M. E. Snodgrass, *World Clothing and Fashion: An Encyclopedia of History, Culture, and Social Influence* (London: Routledge, 2019); J. Morley, *History of Furniture: Twenty-Five Centuries of Style and Design in the Western Tradition* (Boston: Bulfinch Press, 1999).

2 A. Muzquiz, "Flipping through the history of the flip-flop," Heddels [website], April 18, 2018, https://www.heddels.com/2018/04/history-flip-flop/.

3 T. Chenet et al., "Lower limb muscle co-contraction and joint loading of flip-flops walking in male wearers," *PLoS ONE* 13/3 (2018), e0193653; M. Laliberte, "11 reasons why you should never wear flip-flops," The Healthy, April 1, 2021, https://www.thehealthy.com/foot-care/flip-flops-bad-for-feet.

4 We now have a surfeit of general ergonomic books. Notable recent additions include P. M. Bus, *Ergonomics: Foundational Principles, Applications and Technologies* (Boca Raton, FL: CRC Press, 2012); R. Bridger, *Introduction to Human Factors and Ergonomics* (Boca Raton, FL: CRC Press, 2017); and A. Bhattacharya and J. D. McGlothlin (eds),

Occupational Ergonomics: Theory and Applications (Boca Raton, FL: CRC Press, 2019). On the ergonomics of seating, see R. Lueder and K. Noro (eds), *Hard Facts About Soft Machines: The Ergonomics of Seating* (London: Taylor & Francis, 1994).

5 "Who walked the Camino in 2020?," Follow the Camino [website], January 18, 2021, https://followthecamino.com/en/blog/statistics-who-walked-the-camino-in-2020/.

6 Aeroaffaires, "Europe's 20 biggest airports," https://aeroaffaires.com/europes-20-biggest-airports/ [accessed 2021].

7 US Census Bureau, "Census bureau estimates show average one-way travel time to work rises to all-time high," Press release, March 18, 2021, https://www.census.gov/newsroom/press-releases/2021/one-way-travel-time-to-work-rises.html.

8 For a systematic examination of the history of growth of human artifacts and systems, see Smil, *Growth*.

9 N. Goren-Inbar and S. Gonen, *Axe Age: Acheulian Tool-making from Quarry to Discard* (London: Routledge, 2006).

10 Doug Engelbart Institute, "Historic Firsts: 'Father of the Mouse,'" https://www.dougengelbart.org/content/view/162/000/ [accessed 2021].

11 Obviously, not an easy precept to follow after a century of automobilization. Here is one possible approach: J. Speck, *Walkable City* (New York: North Point Press, 2013).

12 L. Appolloni et al., "Walkable urban environments: An ergonomic approach of evaluation," *Sustainability* 12 (2020).

13 World Bank, "Urban population (% of total population)," https://data.worldbank.org/indicator/SP.URB.TOTL.IN.ZS [accessed 2021].

14 New York's Central Park remains a prime example of remarkable planning foresight: R. Rosenzweig and E. Blackmar, *The Park and the People: A History of Central Park* (Ithaca, NY: Cornell University Press, 1998). For crown sizes and space requirements of commonly planted urban trees, see: H. Pretzsch et al., "Crown size and growing space requirement of common tree species in urban centres, parks, and forests," *Urban Forestry & Urban Greening* 14 (2015), pp. 466–79.

15 Hong Kong Housing Authority, "Public housing development in Hong Kong," https://www.housingauthority.gov.hk/en/about-us/

photos-and-videos/videos/public-housing-development-in-hong-kong /index.html [accessed 2020].

16 The most common internal door size in the US is 80 × 36 inches (203.2 × 91.4 centimeters); in England it is 198.1 × 76.2 centimeters.

17 J. J. Shea, *Stone Tools in Human Evolution: Behavioral Differences among Technological Primates* (Cambridge: Cambridge University Press, 2016).

18 For the history of animal harness, see Smil, *Energy and Civilization*, pp. 66–76, 100–3.

19 The longest repeated train trips were the annual pre-1914 migrations of entire families of rich Russians from St. Petersburg and Moscow to France and Italy.

20 W. Jastrzębowski, "Rys Ergonomii czyli Nauki o Pracy opartej na naukach zaczerpniętych z Nauki Przyrody," *Przyroda I Przemysl* (1857), pp. 29–32.

21 F. W. Taylor, *The Principles of Scientific Management* (New York: Harper & Brothers, 1911).

22 C. Barlow et al., "Association between sitting time and cardiometabolic risk factors after adjustment for cardiorespiratory fitness, Cooper Center Longitudinal Study, 2010–2013," *CME ACTIVITY* 13 (December 2016), https://www.cdc.gov/pcd/issues/2016/16_0263.htm.

23 P. Vink et al., "Possibilities to improve the aircraft interior comfort experience," *Applied Ergonomics* 43 (2012), article 354e359.

24 F. L. Smith and B. Cox, "Airline deregulation," Econlib, 2002, https://www.econlib.org/library/Enc/AirlineDeregulation.html.

25 G. Brundrett, "Comfort and health in commercial aircraft: A literature review," *The Journal of the Royal Society for the Promotion of Health* 121 (2001), pp. 29–37.

26 International Civil Aviation Organization, "The World Air Transport in 2019," Annual report (2019), https://www.icao.int/annual-report-2019/Pages/the-world-of-air-transport-in-2019.aspx.

27 Smith and Cox, "Airline deregulation."

28 Seat Guru, "Short-haul economy class comparison chart," https://www.seatguru.com/charts/shorthaul_economy.php [accessed 2021].

29 J. Molenbroek et al., "Thirty years of anthropometric changes relevant to the width and depth of transportation seating spaces, present

and future," *Applied Ergonomics: Human Factors in Technology and Society* 65 (2017), pp. 130–8.

30 Seat Guru, "Short-haul economy class comparison chart."

31 Trust for America's Health, *The State of Obesity: 2020* (Washington, DC: Trust for America's Health, 2020).

32 Center for Engineering & Occupational Safety and Health, *Bariatric Safe Patient Handling and Mobility Guidebook: A Resource Guide for Care of Persons of Size* (St. Louis, MI: CEOSH, 2015).

33 Preferred Health Choice, "Bariatric lift equipment," https://www .phc-online.com/Bariatric_Patient_Lifts_s/49.htm [accessed 2022].

34 World Health Organization, "Obesity and overweight," https://www .who.int/news-room/fact-sheets/detail/obesity-and-overweight [accessed 2021]; Global Obesity Observatory.

35 J. Porta et al., "The ergonomics of airplane seats: The problem with economy class," *International Journal of Industrial Ergonomics* 69 (2019), pp. 90–5.

36 Seat Guru, "Find your seat map," https://www.seatguru.com [accessed 2021].

37 K. Paul, "FAA declines to put a stop to the 'incredible shrinking airline seat,'" Market Watch, July 9, 2018, https://www.marketwatch.com/ story/faa-declines-to-put-a-stop-to-the-incredible-shrinking-airline-seat-2018-07-09.

38 B. Guering, "Seating device comprising a forward-foldable backrest," US Patent application, June 12, 2014, https://loyaltylobby.com/wp-content/uploads/2014/07/Airbus-Seat-Patent-Application.pdf.

39 M. Lane, "Are standing seats a standing joke?," BBC News, July 2, 2010, http://news.bbc.co.uk/2/hi/8779388.stm.

40 N. Hitti, "Layer's smart Move seating for Airbus adapts to the passengers' needs," *dezeen*, February 18, 2019, https://www.dezeen.com/2019/02/18/ layer-move-smart-seating-airbus-economy-technology/.

41 United Nations, *Growth of the World's Urban and Rural Population, 1920–2000* (New York: UN, 1969).

42 G. Alfani and C. Ó Gráda (eds), *Famine in European History* (Cambridge: Cambridge University Press, 2017); A. B. Jannetta, "Famine mortality in nineteenth-century Japan: The evidence from a temple death register," *Population Studies* 46 (1992), pp. 427–43.

43 R. Houston, "Literacy and society in the West, 1500–1850," *Social History* 8 (1983), pp. 269–93.

44 Smil, *Growth*, p. 445.

45 Smil, *Grand Transitions*, p. 154.

46 Smil, *Energy and Civilization*, p. 185.

47 V. Smil, *Creating the Twentieth Century* (New York: Oxford University Press, 2005).

48 There are many comprehensive reviews of this vast topic, from Singer's multivolume classic to Wei's brief account: C. Singer et al. (eds), *A History of Technology*, 5 volumes (Oxford: Oxford University Press, 1954–1958); B. H. Bunch and A. Hellemans, *The Timetables of Technology: A Chronology of the Most Important People and Events in the History of Technology* (New York: Simon & Schuster, 1993); M. Finniston et al., *Oxford Illustrated Encyclopedia of Invention and Technology* (Oxford: Oxford University Press, 1992); J. Wei, *Great Inventions That Changed the World* (Hoboken, NJ: John Wiley, 2012).

49 Smil, *Energy and Civilization*, pp. 146–57.

50 For a complete list of technical specifications of modern vehicles see the ultimateSPECS database: https://www.ultimatespecs.com.

51 For details on the evolution of diesel engines and gas turbines, see Smil, *Prime Movers of Globalization*.

52 L. S. Langston, "Bright future," *Mechanical Engineering* 143/4 (2021), pp. 146–51.

53 S. Lebergott, "Labor Force and Employment, 1800–1960," in D. S. Brady (ed.), *Output, Employment, and Productivity in the United States After 1800* (Cambridge, MA: NBER, 1966), pp. 7–204; FRED, "Current employment statistics (establishment survey)," https://fred.stlouisfed .org/categories/11 [accessed 2020].

54 John Deere, "Tractors," https://www.deere.com/en/tractors/ [accessed 2021].

55 Smil, *Still the Iron Age*.

56 V. Smil, *Enriching the Earth: Fritz Haber, Carl Bosch and the Transformation of World Food Production* (Cambridge, MA: MIT Press, 2001); thyssenkrupp Industrial Solutions, "Making the world's largest ammonia plant even larger," 2019, https://insights.thyssenkrupp-industrial-solutions .com/story/making-the-worlds-largest-ammonia-plant-even-larger/.

57 Long-term shifts in national food consumption can be followed at: Food and Agriculture Organization of the United Nations, FAOSTAT Food Balances, http://www.fao.org/faostat/en/#data [accessed 2021].

58 C. Ingraham, "The absurdity of women's clothing sizes, in one chart," *Washington Post*, August 11, 2015, https://www.washingtonpost.com/news/wonk/wp/2015/08/11/the-absurdity-of-womens-clothing-sizes-in-one-chart/.

59 US Census Bureau, *Historical Statistics of the United States: Colonial Times to 1970* (Washington, DC: USBC, 1975).

60 M. Wallis, *Route 66* (New York: St. Martin's Griffin, 2001).

61 S. Watanabe, "China to expand highway network nearly 50% by 2035," Nikkei Asia, March 16, 2021, https://asia.nikkei.com/Economy/China-to-expand-highway-network-nearly-50-by-2035.

62 Interested readers have a rich choice of general works on growth, ranging from the classic work by D'Arcy Wentworth Thompson, *On Growth and Form* (published in 1917, revised and expanded edition published in 1942 and still in print), to Samuel Brody's detailed examination of growth in animals (1945), to a systematic review of growth phenomena by Robert Banks (1994), to Geoffrey West's explanations of the universal laws of scaling, and to my wide-ranging analysis of growth in all of its organic and anthropogenic forms: D. W. Thompson, *On Growth and Form* (Cambridge: Cambridge University Press, 1917); S. Brody, *Bioenergetics and Growth* (New York: Reinhold, 1945); R. B. Banks, *Growth and Diffusion Phenomena: Mathematical Frameworks and Applications* (Berlin: Springer; 1994); Smil, *Growth*.

63 L. Casson, *Ships and Seamanship in the Ancient World* (Baltimore: Johns Hopkins University Press, 1995).

64 Evergreen Aviation & Space Museum, "Spruce Goose," https://www.evergreenmuseum.org/the-spruce-goose [accessed 2021].

65 G. M. Simons, *The Airbus A380: A History* (Barnsley: Pen and Sword, 2014).

66 M. Bowman, *Boeing 747: A History: Delivering the Dream* (Oxford: Casemate, 2021); W. Scheller, *Ford F-Series: America's Pickup Truck* (New York: Universe, 2008); M. Boas and S. Chaim, *Big Mac: The Unauthorized Story of McDonald's* (New York: Dutton, 1976).

67 W. Yang et al., "On the tear resistance of skin," *Nature Communications* 6 (2015), p. 6649.

68 C. J. Williams et al., "Helically arranged cross struts in azhdarchid pterosaur cervical vertebrae and their biomechanical implications," *iScience* (2021), 102338.

69 M. P. Witton, "Titans of the skies: Azhdarchid pterosaurs," *Geology Today* 23 (2007), pp. 33–8.

70 A. Carder, *Forest Giants of the World: Past and Present* (Markham, ON: Fitzhenry & Whiteside, 1995)

71 K. J. Niklas, "Maximum plant height and the biophysical factors that limit it," *Tree Physiology* 27 (2007), pp. 433–40.

72 G. Galilei, *Discorsi e dimostrazioni matematiche intorno a due nuove scienze* (Leiden: Elsevier, 1638). English translation (with original illustrations) is readily available at: http://galileoandeinstein.physics.virginia.edu/ tns_draft/index.htm.

73 Engineering Toolbox, "Young's modulus – tensile and yield strength for common materials," 2003, https://www.engineeringtoolbox.com/ young-modulus-d_417.html.

74 N. Berg, "Is there a limit to how tall buildings can get?," Bloomberg, August 16, 2012, https://www.bloomberg.com/news/articles/2012- 08-16/is-there-a-limit-to-how-tall-buildings-can-get.

75 British Petroleum, *Statistical Review of World Energy* (London: BP, 2021).

76 Vessel Tracking, "Seawise Giant—the biggest ship ever built," https:// www.vesseltracking.net/article/seawise-giant [accessed 2021].

77 Raunek, "The ultimate guide to ship sizes," Marine Insight, February 1, 2021, https://www.marineinsight.com/types-of-ships/the-ultimate- guide-to-ship-sizes/.

78 S. Mambra, "The complete story of the EXXON Valdez oil spill," Marine Insight, March 23, 2022, https://www.marineinsight.com/ maritime-history/the-complete-story-of-the-exxon-valdez-oil-spill/.

79 Raunek, "Alang, Gujarat: The world's biggest ship breaking yard & a dangerous environmental time bomb," Marine Insight, March 11, 2021, https://www.marineinsight.com/environment/alang-gujarat- the-world's-biggest-ship-breaking-yard-a-dangerous-environmental- time-bomb/.

80 Zeymarine, "Largest oil tankers ever built," November 17, 2020, https://zeymarine.com/largest-oil-tankers-ever-built/.

81 GE Renewable Energy, "Haliade-X offshore wind turbine," 2021, https://www.ge.com/renewableenergy/wind-energy/offshore-wind/haliade-x-offshore-turbine.

82 E. Loth et al., "Downwind pre-aligned rotors for extreme-scale wind turbines," *Wind Energy* 20 (2017), pp. 1241–59.

83 G. Moore, "Cramming more components onto integrated circuits," *Electronics* 38/8 (1965), pp. 114–17.

84 M. Roser and H. Ritchie, "Moore's Law: The number of transistors on microchips doubles every two years," Our World in Data, 2021 (revised March 2022), https://ourworldindata.org/technological-change.

85 Car manufacturing was expected to be the fastest-growing market for microchips between 2017 and 2021: A. Lawrence and J. VerWey, "The automotive semiconductor market—key determinants of U.S. firm competitiveness," Executive briefing on trade, May 2019, https://www.usitc.gov/publications/332/executive_briefings/ebot_amanda_lawrence_john_verwey_the_automotive_semiconductor_market_pdf.pdf.

86 M. Feldman, "Dennard scaling demise puts permanent dent in supercomputing," The Next Platform, June 18, 2019, https://www.nextplatform.com/2019/06/18/dennard-scaling-demise-puts-permanent-dent-in-supercomputing/.

87 R. H. Dennard et al., "Design of ion-implanted MOSFET's with very small physical dimensions," *IEEE Journal of Solid-State Circuits* 9 (1974), pp. 256–68.

88 National Research Council, *The Future of Computing Performance* (Washington, DC: The National Academies Press, 2011).

89 D. Black, "10nm, 7nm, 5nm . . . Should the chip nanometer metric be replaced?," HPC Wire, June 1, 2020, https://www.hpcwire.com/2020/06/01/10nm-7nm-5nm-should-the-chip-nanometer-metric-be-replaced/.

90 L. Dormehl, "Computers can't keep shrinking, but they'll keep getting better," Digital Trends, March 17, 2018, https://www.digitaltrends.com/computing/end-moores-law-end-of-computers/.

91 R. Johnson, "GlobalFoundries stops development of its 7-nm LP node," Tech Report, August 27, 2018, https://techreport.com/news/34033/globalfoundries-stops-development-of-its-7-nm-lp-node.

V SIZE AND SCALING

1 Everything you ever wanted to know about lions is summarized in publications listed in Craig Packer's review: C. Packer, "The African lion: A long history of interdisciplinary research," *Frontiers in Ecology and Evolution* 7 (2019), p. 259.

2 Swift, *Gulliver's Travels*. Many e texts are available, including: https://www.fulltextarchive.com/page/Gulliver-s-Travels/.

3 J. Boswell, *The Life of Johnson* (Oxford: Oxford University Press, 1969), p. 595.

4 In a letter to Alexander Pope on September 29, 1725, Swift wrote about the book's intent: "the chief end I propose to myself in all my labours, is to vex the world rather than divert it." See T. Sheridan et al. (eds), *The Works of the Rev. Jonathan Swift*, vol. 14 (London: J. Johnson, 1803), p. 37.

5 G. Galofré-Vilà et al., "Heights Across the Last 2000 Years in England," Oxford Economic and Social History Working Papers, University of Oxford, Department of Economics (2017).

6 CDC, "Adult BMI calculator," https://www.cdc.gov/healthyweight/assessing/bmi/adult_bmi/metric_bmi_calculator/bmi_calculator.html [accessed 2021].

7 F. Moog, "Gulliver was a bad biologist," *Scientific American* 18/5 (1948), pp. 52–6. Two other notable corrections of Swift's errors are: L. R. Moreira de Carvalho, "Gulliver, os liliputianos e a física," *Revista Brasileira de Ensino de Física* 36 (2014); T. Kuroki, "Physiological essay on *Gulliver's Travels*: a correction after three centuries," *The Journal of Physiological Sciences* 69 (2019), pp. 421–4. Kuroki was not aware of either Moog's or Moreira de Carvalho's publications.

8 A. Quetelet, *Sur l'homme et sur le développement de ses facultés: Essai de physique sociale* (Paris: Bachelier, 1835), p. 52.

9 "Lilliput and Blefescu," The Science of Gulliver's Travels [website], http://thescienceofgulliverstravels.weebly.com/lilliput-and-blefescu.html.

10 A. Keys et al., "Indices of relative weight and obesity," *Journal of Chronic Diseases* 25 (1972), pp. 329–43.

11 New England Primate Conservancy, "Cotton-top tamarin," https://neprimateconservancy.org/cotton-top-tamarin/ [accessed 2021].

12 Even the most extreme (and highly uncertain) estimates credit the extinct *Gigantopithecus* with a height of less than 3 meters.

13 Swift in his last birthday poem to Stella (Esther Johnson), written on March 13, 1727 (she died less than a year later): "From not the gravest of divines / Accept for once some serious lines."

14 G. Galilei, *Discorsi e dimostrazioni matematiche intorno a due nuove scienze* (Leiden: Elsevier, 1638). English translation with original illustrations is readily available at: http://galileoandeinstein.physics.virginia.edu/ tns_draft/index.htm.

15 P. Morrison, "Scaling—the physics of Lilliput," 1968, https://lru.praxis .dk/Lru/microsites/hvadermatematik/hem1download/kap5_Projekt _5.9_Scaling_the_Physics_of_Lilliput.pdf.

16 National Museum of Natural History, "African bush elephant," https:// naturalhistory.si.edu/exhibits/african-bush-elephant [accessed 2021].

17 All estimates of the largest sauropod masses, based on very limited skeletal remains, are highly uncertain; the latest range for *Argentino- saurus* is 65–75 tons: G. Paul, "Determining the largest known land animal: A critical comparison of differing methods for restoring the volume and mass of extinct animals," *Annals of Carnegie Museum* 85 (2019), pp. 335–58.

18 J. B. S. Haldane, "On Being the Right Size," in *Possible World and Other Essays* (London: Chatto and Windus, 1926), pp. 18–27.

19 The Marine Mammal Center, "Blue whale," https://www.marine mammalcenter.org/animal-care/learn-about-marine-mammals/cetaceans/ blue-whale [accessed 2021].

20 T. A. Lumpkin, "How a Gene from Japan Revolutionized the World of Wheat: CIMMYT's Quest for Combining Genes to Mitigate Threats to Global Food Security," in Y. Ogihara et al. (eds), *Advances in Wheat Genetics: From Genome to Field* (Berlin: Springer-Verlag, 2015), pp. 13–20; T. Würschum, "A modern Green Revolution gene for reduced height in wheat," *The Plant Journal* 92 (2017), pp. 892–903.

21 C. M. Donald and J. Hamblin, "The biological yield and harvest index of cereals as agronomic and plant breeding criteria," *Advances in Agron- omy* 28 (1976), pp. 361–405.

22 Iowa Corn, "Corn FAQs," https://www.iowacorn.org/education/ faqs [accessed 2022].

23 North Dakota State University, "Optimal seeding rates," https://www.ag.ndsu.edu/crops/spring-wheat-articles/optimal-seeding-rates [accessed 2021].

24 Guadua Bamboo, "How to plant bamboo," https://www.guadua bamboo.com/blog/how-to-plant-bamboo [accessed 2022].

25 K. Meeh, "Oberflächenmessungen des menschlichen Körpers," *Zeitschrift für Biologie* 15 (1879), pp. 425 8.

26 D. DuBois and E. F. DuBois, "The measurement of the surface area of man," *Archives of Internal Medicine* 16 (1915), pp. 868–81.

27 O. Snell, "Die Abhängigkeit des Hirngewichts von dem Körpergewicht und den geistigen Fähigkeiten," *Archiv für Psychiatrie und Nervenkrankenheiten* 23 (1891), pp. 436–46.

28 E. Dubois, "Sur le rapport de l'encéphale avec la grandeur du corps chez les Mammifères," *Bulletin Société d'Anthropologie Paris*, 4e série, 8 (1897), pp. 337–74.

29 L. Lapicque, "Tableau général des poids somatiques et encéphaliques dans les espèces animales," *Bulletin Société d'Anthropologie Paris*, 5e série, 9 (1907), pp. 248–69.

30 J. S. Huxley, "Constant differential growth-ratios and their significance," *Nature* 114 (1924), pp. 895–6.

31 J. S. Huxley, *Problems of Relative Growth* (London: Methuen & Company, 1932).

32 J. S. Huxley and G. Teissier, "Terminology of relative growth rates," *Nature* 137 (1936), pp. 780–1.

33 J. Needham and I. M. Lerner, "Terminology of relative growth," *Nature* 146 (1940), p. 618.

34 M. Kleiber, "Body size and metabolism," *Hilgardia* 6/11 (1932), pp. 315–53.

35 G. Redlarski et al., "Body surface area formulae: an alarming ambiguity," *Anaesthesia* 58 (2003), pp. 50–83.

36 E. Font et al., "Rethinking the effects of body size on the study of brain size evolution," *Brain, Behavior and Evolution* 93 (2019), pp. 182–95.

37 Here is an incomplete list of journals explicitly devoted to the brain, with much more research published in other neuroscience, neurology, physiology, genetic, and medical journals: *Brain*; *Brain and Behavior*; *Brain, Behavior and Evolution*; *Brain, Behavior and Immunity*; *Brain and*

Cognition; *Brain Injury*; *Brain Pathology*; *Brain Research*; *Brain Research Bulletin*; *Brain Stimulation*; *Brain Structure and Function*; *Human Brain Mapping*.

38 D. K. Molina et al., "Normal organ weights in men: Part II—the brain, lungs, liver, spleen and kidneys," *American Journal of Forensic Medical Pathology* 33 (2012), pp. 362–7.

39 F. A. Azevedo et al., "Equal numbers of neuronal and nonneuronal cells make the human brain an isometrically scaled-up primate brain," *Journal of Comparative Neurology* 513 (2009), pp. 532–4.

40 M. Henneberg, "Evolution of the human brain: Is bigger better?," *Clinical and Experimental Pharmacology and Physiology* 25 (2007), pp. 745–9.

41 Recall that brains of improperly scaled Lilliputians would have only 0.06 percent of human synapses.

42 S. L. Lindstedt and P. J. Schaeffer, "Use of allometry in predicting anatomical and physiological parameters of mammals," *Laboratory Animals* 36 (2002), pp. 1–19.

43 E. P. Snelling et al., "Scaling of morphology and ultrastructure of hearts among wild African antelope," *Journal of Experimental Biology* 221 (2018), jeb184713.

44 T. H. Dawson, "Allometric relations and scaling laws for the cardiovascular system of mammals," *Systems* 2 (2014), pp. 168–85.

45 A. R. Casha et al., "Physiological rules for the heart, lungs and other pressure-based organs," *Journal of Thoracic Diseases* 9 (2017), pp. 3793–801.

46 B. Günther and E. Morgado, "Allometric scaling of biological rhythms in mammals," *Biological Research* 38 (2005), pp. 207–12; T. H. Dawson, "Allometric relations and scaling laws for the cardiovascular system of mammals," *Systems* 2 (2014), pp. 168–85.

47 S. Ikeda, "The cardiac surgery training using pig hearts and small pumps," *Journal of Surgical Techniques and Procedures*, 2/1 (2018), p. 1015.

48 R. M. Reed et al., "Cardiac size and sex matching in heart transplantation: Size matters in matters of sex and the heart," *JACC Heart Failure* 2/1 (2014), pp. 73–83.

49 A. Antoł and J. Kozłowski, "Scaling of organ masses in mammals and birds: Phylogenetic signal and implications for metabolic rate scaling," *ZooKeys* 982 (2020), pp. 149–59.

50 H. D. Prange et al., "Scaling of skeletal mass to body mass in birds and mammals," *The American Naturalist* 113 (1979), pp. 103–22.

51 M. N. Muchlinski et al., "Muscle mass scaling in primates: An energetic and ecological perspective," *American Journal of Primatology* 74 (2012), pp. 395–407.

52 I. Janssen et al., "Skeletal muscle mass and distribution in 468 men and women aged 18–88 yrs," *Journal of Applied Physiology* 89 (2012), pp. 81–8.

53 H. C. Howland et al., "The allometry and scaling of the size of vertebrate eyes," *Vision Research* 44 (2004), pp. 2043–65.

54 R. A. Powell and M. S. Mitchell, "What is a home range?," *Journal of Mammalogy* 93 (2012), pp. 948–58.

55 N. Tamburello et al., "Range scale energy and the scaling of animal space use," *The American Naturalist* 186 (2015), pp. 196–211.

56 R. L. Nudds et al., "Evidence for a mass dependent step-change in the scaling of efficiency in terrestrial locomotion," *PLoS ONE* 4/9 (2009), e6927.

VI METABOLIC SCALING

1 Although Swift's correspondents—above all, Alexander Pope and Henry St. John (Viscount Bolingbroke)—were not scientists but men of poetry and politics, they certainly had access to the best English (and French) libraries.

2 F. Sarrus and J. Rameaux, "Rapport sur une mémoire adressée à l'Académie royale de Médecine," *Bulletin Académie Royale de Médecine Paris* 3 (1838), pp. 1094–100.

3 M. Rubner, "Über den Einfluss der Körpergrösse auf Stoff- und Kraftwechsel," *Zeitschrift für Biologie* 19 (1883), pp. 536–62.

4 E. Voit, "Über die Grösse des Energiebedarfs der Tiere im Hungerzustande," *Zeitschrift für Biologie* 41 (1901), pp. 113–54.

5 Kleiber expanded his original analysis (see chapter V, note 34) in 1947: M. Kleiber, "Body size and metabolic rate," *Physiological Reviews* 27 (1947), pp. 511–41; and in 1961: *The Fire of Life: An Introduction to Animal Energetics* (New York: Wiley, 1961).

6 Mathbits, "Line of best fit," https://mathbits.com/MathBits/TISection/Statistics1/LineFit.htm [accessed 2021].

7 FAO/WHO/UNU, *Human Energy Requirements: Report of a Joint FAO/WHO/UNU Expert Consultation* (Rome: FAO, 2004).

8 G. B. West et al., "A general model for the origin of allometric scaling laws in biology," *Science* 276 (1997), pp. 122–6. This was followed by many other publications and eventually by a comprehensive book: G. West, *Scale* (New York: Penguin, 2017).

9 S. E. Rampal et al., "Demystifying the West, Brown & Enquist model of the allometry of metabolism," *Functional Ecology* 20 (2006), pp. 394–9.

10 C.-A. Darveau et al., "Darveau et al. reply," *Nature* 421 (2003), p. 714.

11 H. Li et al., "Lack of evidence for ¾ scaling of metabolism in terrestrial plants," *Journal of Integrative Plant Biology* 47 (2005), pp. 1173–83.

12 C. R. White and R. S. Seymour, "Mammalian basal metabolic rate is proportional to body mass 2/3," *Proceedings of the National Academy of Sciences* 100 (2003), pp. 4046–9.

13 I. Capellini et al., "Phylogeny and metabolic scaling in mammals," *Ecology* 91 (2010), pp. 2783–93.

14 K. R. Westerterp, "Doubly labelled water assessment of energy expenditure: Principle, practice, and promise," *European Journal of Applied Physiology* 117 (2017), pp. 1277–85.

15 K. A. Nagy, "Field metabolic rate and body size," *The Journal of Experimental Biology* 208 (2005), pp. 1621–5.

16 Calculated from data in White and Seymour (see note 12 above).

17 C. R. White et al., "The scaling and temperature dependence of vertebrate metabolism," *Biological Letters* 2/1 (2005), pp. 125–7.

18 C. R. White et al., "Allometric exponents do not support a universal metabolic allometry," *Ecology* 88/2 (2007), pp. 315–23.

19 A. M. Makarieva et al., "Mean mass-specific metabolic rates are strikingly similar across life's major domains: Evidence for life's metabolic optimum," *Proceedings of the National Academy of Sciences* 105 (2008), pp. 16994–9.

20 S. L. Chown et al., "Scaling of insect metabolic rate is inconsistent with the nutrient supply network model," *Functional Ecology* 21 (2007), pp. 282–90; A. J. Riveros and B. J. Enquist, "Metabolic scaling in

insects supports the predictions of the WBE model," *Journal of Insect Physiology* 57 (2011), pp. 688–93.

21 A. I. Bruce and M. Burd, "Allometric scaling of foraging rate with trail dimensions in leaf-cutting ants," *Proceedings of the Royal Society B* 279 (2012), pp. 2442–7.

22 FAO/WHO/UNU, *Human Energy Requirements*.

23 E. M. Widdowson, "How much food does man require?," in J. Mauron (ed.), *Nutritional Adequacy, Nutrient Availability and Needs* (Basel: Birkhäuser Verlag, 1983), pp. 11–25.

24 N. G. Norgan et al., "The energy and nutrient intake and the energy expenditure of 204 New Guinean adults," *Philosophical Transactions of the Royal Society of London: Series B, Biological Sciences* 268 (1974), pp. 309–48.

25 A. M. Prentice, "Adaptations to long-term low energy intake," in *Energy Intake and Activity*, ed. E. Pollitt and P. Amante (New York: Alan R. Liss, 1984), pp. 3–31.

26 C. J. K. Henry and D. C. Rees, "New predictive equations for the estimation of basal metabolic rate in tropical peoples," *European Journal of Clinical Nutrition* 45 (1991), pp. 177–85; L. S. Piers and P. S. Shetty, "Basal metabolic rates of Indian women," *European Journal of Clinical Nutrition* 47 (1993), pp. 586–91.

27 EatPeru, "Eating cuy: Peruvian Guinea pig delicacy," https://www.eatperu.com/eating-cuy-guinea-pig-peruvian-delicacy/ [accessed 2021].

28 Smil, *Prime Movers of Globalization*.

29 S. Takahashi et al., "Development of micro cogeneration system with a porous catalyst microcombustor," *Journal of Physics: Conference Series* 557 (2014); Wärtsilä, "The world's most powerful engine enters service," Press release, 12 September 2006, https://www.wartsila.com/media/news/12-09-2006-the-world%27s-most-powerful-engine-enters-service.

30 V. Smil, "Electric container ships are a hard sail," *IEEE Spectrum* (March 2019), p. 22.

31 DVB Bank SE, *An Overview of Commercial Aircraft 2018–2019* (December 2017), https://www.dvbbank.com/~/media/Files/D/dvbbank-corp/aviation/dvb-overview-of-commercial-aircraft-2018-2019.pdf.

32 T. McMahon and J. T. Bonner, *On Size and Life* (New York: Scientific American Library, 1985).

33 S. Brown et al., "Investigation of scaling laws for combustion engine performance," Oregon State University (2017), https://sites01.lsu.edu/faculty/smenon/wp-content/uploads/sites/133/2017/02/WSSCI_Provo_v5.pdf.

34 R. D. Lorenz, "Flight power scaling of airplanes, airships, and helicopters: Application to planetary exploration," *Journal of Flight* 38 (2001), pp. 208–14.

35 J. H. Marden and L. R. Allen, "Molecules, muscles, and machines: Universal performance characteristics of motors," *Proceedings of the National Academy of Sciences* 99 (2002), pp. 4161–6.

36 Pratt & Whitney, "JT9D engine," https://prattwhitney.com/products-and-services/products/commercial-engines/jt9d [accessed 2021]; GE Aviation, "GE90 Commercial Aircraft Engine," https://www.geaviation.com/propulsion/commercial/ge90 [accessed 2021].

37 United Nations, "Cities: A 'cause of and solution to' climate change," UN News, September 18, 2019, https://news.un.org/en/story/2019/09/1046662.

38 M. Fragkias et al., "Does size matter? Scaling of CO_2 emissions and U.S. urban areas," *PLoS ONE* 8/6 (2013), e64727.

VII SYMMETRIES AROUND MEANS

1 Tall Life, "Height percentile calculator, by age or country," https://tall.life/height-percentile-calculator-age-country/ [accessed 2022]; NCD Risk Factor Collaboration, "A century of trends in adult human height," *eLife* 5 (2016), e13410.

2 Perhaps the best country to check uniform, and uniformed, heights of children is Japan. Sunflower fields are not as common in the EU and North America as they used to be, displaced by mass-scale cultivations of rapeseed: Ukraine, Russia, Argentina, and China remain their large growers.

3 The blue jay's body length (from the tip of its bill to the tip of its tail) is up to 30 centimeters; its wingspan ranges between 34 and 43 centimeters. Doubling that wingspan would match the size of a peregrine

falcon. Canadian Wildlife Federation, "Blue jay," https://www.hww
.ca/en/wildlife/birds/blue-jay.html [accessed 2021].

4   A. M. Stigler, *The History of Statistics* (Cambridge, MA: Belknap Press,
1986).

5   S. Stahl, "The evolution of the normal distribution," *Mathematics
Magazine* 79 (2006), pp. 96–113.

6   A. Hald, "Galileo's statistical analysis of astronomical observations,"
International Statistical Review 54 (1986), pp. 211–20.

7   A. de Moivre, *Approximatio ad Summam Terminorum Binomii (a+b)n in
Seriem Expansi* (Printed for private circulation, 1733).

8   P.-S. Laplace, "Mémoire sur la probabilité des causes par les évène-
ments," *Mémoires de l'Academie royale des sciences presentés par divers savans*
6 (1774), pp. 621–56; P.-S. Laplace, *Théorie Analytique des Probabilités*
(Paris: Courcier, 1812); C. F. Gauss, *Theoria Motus Corporum Celestium*
(Hamburg: Perthes et Besser, 1809). [Translated by C. H. Davis as:
*Theory of Motion of the Heavenly Bodies Moving about the Sun in Conic Sec-
tions* (Boston: Little, Brown 1857).]

9   A. Quetelet, *Sur l'homme et le Développement de ses Facultés, ou Essai de
Physique Social* (Paris: Bacheller, 1835).

10 A. Quetelet, *Letters Addressed to H.R.H. the Grand Duke of Saxe Coburg
and Gotha, on the Theory of Probabilities as Applied to the Moral and Political
Sciences* (London: Charles & Edwin Layton, 1842).

11 F. Galton, *Natural Inheritance* (London and New York: Macmillan,
1889), p. 66.

12 Description of the model is in Galton's just-cited book, on pp. 63–5.

13 Sold under three different names: Qincunx, Dalton board, and
Binostat Probability Demonstrator Game.

14 R. Pearl, *Introduction to Medical Biometry and Statistics* (Philadelphia, PA:
W. B. Saunders, 1923); R.E. Chaddock, *Principles and Methods of Statistics* (Boston: Houghton Mifflin, 1925).

15 Here are just two of many choices: "Standard normal distribution
table,"https://www.soa.org/globalassets/assets/Files/Edu/2018/exam-
srm-tables.pdf; Engineering Statistics Handbook, "Cumulative distribution function of the standard normal distribution," https://www.itl
.nist.gov/div898/handbook/eda/section3/eda3671.htm [accessed 2022].

16 For example: https://www.mathsisfun.com/data/standard-normal-distribution-table.html.

17 A. Dance, "Survival of the littlest: The long-term impacts of being born extremely early," *Nature* 582 (2020), pp. 20–3.

18 A. Carder, *Forest Giants of the World: Past and Present* (Markham, ON: Fitzhenry & Whiteside, 1995).

19 National Museum of Natural History, "African Bush Elephant: Meet Henry," https://naturalhistory.si.edu/exhibits/african-bush-elephant [accessed 2021].

20 Audubon Guide to North American Birds, "Anna's hummingbird," https://www.audubon.org/field-guide/bird/annas-hummingbird [accessed 2021].

21 N. Page, *Lord Minimus: The Extraordinary Life of Britain's Smallest Man* (New York: St. Martin's Press, 2002).

22 E. Ragusa (ed.), *Velázquez* (New York: Rizzoli, 2004).

23 P. Carlson, "Rendezvous: P. T. Barnum Meets Queen Victoria," History Net, March 21, 2018, https://www.historynet.com/rendezvous-p-t-barnum-meets-queen-victoria/.

24 For Barnum's career seen from a broader cultural perspective: B. Adams, *E Pluribus Barnum: The Great Showman and the Making of U.S. Popular Culture* (Minneapolis: University of Minnesota Press, 1997).

25 Mensa International, https://www.mensa.org [accessed 2021].

26 NCAA Men's D1 Basketball, https://www.ncaa.com/sports/basketball-men/d1; NBA, https://www.nba.com [accessed 2021].

27 I. Zarić et al., "Body height of elite basketball players: Do taller basketball teams rank better at the FIBA World Cup?," *International Journal of Environmental Research and Public Health* 17 (2020), p. 3141.

28 Scholarship Stats, "Basketball demographics: NCAA I basketball players by height," https://scholarshipstats.com/NCAA1basketball [accessed 2021].

29 D. Curcic, "69 Years of Height Evolution in the NBA [4,379 players analyzed]," Run Repeat, https://runrepeat.com/height-evolution-in-the-nba [accessed 2021].

30 In 2020, the average NBA height was 6'6", the lowest during the past 40 years.

31 A. Sedeaud et al., "Secular trend: Morphology and performance," *Journal of Sports Sciences* 32/12 (2014), pp. 1146–54.

32 A. J. Wilcox, "On the importance—and the unimportance—of birthweight," *International Journal of Epidemiology* 30 (2001), pp. 1233–41; R. Charnigo et al., "Thinking outside the curve, part I: modeling birthweight distribution," *BMC Pregnancy and Childbirth* 10 (2010), p. 37.

33 J. Jeon et al., "Optimal birth weight and term mortality risk differ among different ethnic groups in the U.S.," *Scientific Reports* 9 (2019), p. 1651.

34 S. Vangen et al., "The heavier the better? Birthweight and perinatal mortality in different ethnic groups," *International Journal of Epidemiology* 31 (2002), pp. 654–60.

35 CDC, "Birth to 36 months: Girls," May 30, 2000, https://www.cdc .gov/growthcharts/data/set1clinical/cj41l018.pdf; WHO, "Length/ height-for-age," https://www.who.int/tools/child-growth-standards/ standards/length-height-for-age [accessed 2021].

36 CDC, "2 to 20 years: Girls," May 30, 2000, https://www.cdc .gov/growthcharts/data/set2clinical/cj41l072.pdf.

37 A. G. Warrener et al., "A wider pelvis does not increase locomotor cost in humans, with implications for the evolution of childbirth," *PLoS ONE* 10/3 (2015), e0118903; L. Betti and A. Manica, "Human variation in the shape of the birth canal is significant and geographically structured," *Proceedings of the Royal Society B* 285 (2018), article 20181807.

38 M. Lipschuetz et al., "A large head circumference is more strongly associated with unplanned cesarean or instrumental delivery and neonatal complications than high birthweight," *American Journal of Obstetrics & Gynecology* 213/6 (2015), 833.e1–e12.

39 Height percentile calculators by gender, age and country are available at https://tall.life/height-percentile-calculator-age-country/.

40 S. Bestul, "How to score a buck—the right way," Field and Stream, December 2, 2019, https://www.fieldandstream.com/story/hunting/ how-to-score-buck-antlers-the-right-way/.

41 S. W. Stedman, *The Bell Curve of Mature Buck Antler Scores: When You Manage a Buck Herd, You Manage a Bell Curve of Antler Scores* (Caesar Kleberg Wildlife Research Institute, 2016), https://www.ckwri

.tamuk.edu/sites/default/files/pdf-attachment/2016-05/bell_curve
_series_part_i_final.pdf.

42 C. W. Adams et al., *Six Sigma Deployment* (Burlington, MA: Butterworth-Heinemann, 2003); "DPMO: Sigma Level Table," http://www
.sixsigmadigest.com/support-files/DPMO-Sigma-Table.pdf.

43 Coca-Cola Great Britain, "How many drinks does The Coca-Cola Company sell worldwide each day?," October 1, 2020, https://www.coca-cola
.co.uk/our-business/faqs/how-many-cans-of-coca-cola-are-sold-worldwide-in-a-day.

44 W. R. MacDonell, "On criminal anthropometry and the identification of criminals," *Biometrika* 1 (1902), pp. 177–227.

45 Student, "The probable error of a mean," *Biometrika* 6 (1908), pp. 1–25.

VIII WHEN ASYMMETRIES RULE

1 That record is now held by Nirmal Purja. who climbed the 14 peaks in just six months and six days: J. Daley, "Nepalese mountaineer smashes speed record for climbing world's 14 tallest peaks," *Smithsonian*, October 30, 2019, https://www.smithsonianmag.com/smart-news/nepalese-mountaineer-smashes-speed-record-climbing-worlds-14-tallest-peaks-180973437/.

2 Misi-Ziibi in Anishinaabemowin (Ojibwe): The Decolonial Atlas, "Native names for the Mississippi River," January 5, 2015, https://decolonialatlas
.wordpress.com/2015/01/05/native-names-for-the-mississippi-river/.

3 J. Korčák, "Deux types fondamentaux de distribution statistique," *Bulletin de l'Institut Internationale de Statistique* 3 (1938), pp. 295–9. Much expanded Czech version: J. Korčák, "Přírodní dualita statistického rozložení," *Statistický Obzor* 22/5–6 (1938), pp. 171–222.

4 J. Hemming, *Tree of Rivers: The Story of the Amazon* (London: Thames & Hudson, 2008).

5 US Geological Survey, "Cool earthquake facts," https://www.usgs.gov/programs/earthquake-hazards/cool-earthquake-facts [accessed 2021].

6 US Geological Survey, "The 2019 Ridgecrest, California, earthquake sequence," https://earthquake.usgs.gov/storymap/index-ridgecrest
.html [accessed 2020].

7   R. B. Stothers, "The Great Tambora eruption in 1815 and its after-math," *Science* 224 (1984), pp. 1191–8.

8   Office of National Statistics, "Average household income, UK: financial year 2020," January 21, 2021, https://www.ons.gov.uk/peoplepopulation andcommunity/personalandhouseholdfinances/incomeandwealth /bulletins/householddisposableincomeandinequality/financialyear2020.

9   Counting the world's billionaires remains, to a large degree, a matter of best estimates rather than one of accurate accounts (they do not share the details of their portfolios with the editors of *Forbes* maga-zine), and annual shifts in the ranking are often surprisingly large.

10 CDC, "2009 H1N1 Pandemic (H1N1pdm09 virus)," https://www .cdc.gov/flu/pandemic-resources/2009-h1n1-pandemic.html [accessed 2019]; J. M. Barry, *The Great Influenza* (New York: Penguin, 2005).

11 F. Galton, "The geometric mean, in vital and social sciences," *Proceed-ings of the Royal Society* 29–30 (1879), p. 367.

12 C. G. A. Harrison et al., "Continental hypsography," *Tectonics* 2 (1983), pp. 357–77; Wikimedia Commons, "Hypsometric curves," https:// commons.wikimedia.org/wiki/Category:Hypsometric_curves.

13 P. Jackson, *The Mongols and the West, 1221–1410* (London: Routledge, 2018).

14 US Census Bureau, "City and town population totals: 2010–2019," https://www.census.gov/data/tables/time-series/demo/popest/2010s-total-cities-and-towns.html [accessed 2019].

15 S. Newcomb, "Note on the frequency of use of the different digits in natural numbers," *American Journal of Mathematics* 4 (1881), pp. 39–40.

16 F. Benford, "The law of anomalous numbers," *Proceedings of the American Philosophical Society* 78 (1938), pp. 551–72.

17 V. Pareto, *Cours d'Économie Politique Professé a l'Université de Lausanne*, vol. II (Lausanne: Rouge, 1896), p. 312.

18 F. Auerbach, "Das Gesetz der Bevölkerungskonzentration," *Petermanns Geographische Mitteilungen* 59 (1913), pp. 73–6; J. Estoup, *Les Gammes Sténographiques* (Paris: Gauthier-Villars, 1916).

19 U. G. Yule, "A mathematical theory of evolution, based on the conclu-sions of Dr. J. C. Willis, F.R.S.," *Philosophical Transactions of the Royal Society of London. Series B* 213 (1925), pp. 21–87; A. J. Lotka, "The fre-quency distribution of scientific productivity," *Journal of the Washington Academy of Sciences* 16 (1926), pp. 317–24.

20 In all of these cases Korčák attempted to gather the greatest possible variety of measurements, but these were largely limited to European phenomena.

21 C. Frohlich, "Kiyo Wadati and early research on deep focus earthquakes," *Journal of Geophysical Research* 92 (1987), pp. 777–88; B. Gutenberg and C. F. Richter. "Frequency of earthquakes in California," *Bulletin of the Seismic Society of America* 34 (1944), pp. 185–8.

22 L. F. Richardson, "Variation of the frequency of fatal quarrels with magnitude," *Journal of the American Statistical Association* 43 (1948), pp. 523–46.

23 G. K. Zipf, *The Psycho-Biology of Language* (Cambridge, MA: Harvard University Press, 1935).

24 G. K. Zipf, *Human Behavior and the Principle of Least Effort* (Boston: Addison-Wesley Press, 1949).

25 D. A. Seekell and M. L. Pace, "Does the Pareto distribution adequately describe the size-distribution of lakes?," *Limnology and Oceanography* 56 (2011), pp. 350–6; D. Sornette et al., "Rank-ordering statistics of extreme events: Application to the distribution of large earthquakes," *Journal of Geographical Research* 101 (1996), pp. 13883–93; F. Meng et al., "Power law relations in earthquakes from microscopic to macroscopic scales," *Scientific Reports* 9 (2019), 10705.

26 B. D. Malamud et al., "Characterizing wildfire regimes in the United States," *Proceedings of the National Academy of Sciences* 29 (2005), pp. 4694–9; D. L. Turcotte and B. D. Malamud, "Landslides, forest fires and earthquakes: Examples of self-organized critical behavior," *Physica A* 340 (2004), pp. 580–9; C. Verbeeck et al., "Solar flare distributions: Lognormal instead of power law?," *The Astrophysical Journal* 884 (2019), p. 50.

27 H. D. Rosenfled et al., "The area and population of cities: New insights from a different perspective on cities," *American Economic Review* 101 (2011), pp. 2205–25; K. Giesen and J. Suedekum, *The Size Distribution Across all "Cities": A Unifying Approach* (Hamburg: Institute for the Study of Labor, 2012); R. Gonzalez-Val et al., "Size distributions for all cities: Which one is best?," *Papers in Regional Science* 94 (2013).

28 Just a few outstanding selections: M. Levy and S. Solomon, "New evidence for the power-law distribution of wealth," *Physica A* 242 (1997), pp. 90–4; K. Okuyama et al., "Zipf's law in income distribution of

companies," (1999), pp. 125–31; R. L. Axtell, "Zipf distribution of U.S. firm sizes," *Science* 293 (2001), pp. 1818–20; A. Ishikawa et al., "A new approach to personal income distribution," *Physica A* 334 (2002), pp. 255–66; Y. Fujiwara et al., "Growth and fluctuations of personal income," *Physica A* 321 (2003), pp. 598–604; S. Aoki, *Pareto Distributions and the Evolution of Top Incomes in the U.S.*, MPRA Paper 47967 (Munich: University Library of Munich, 2013).

29 A. B. Godfrey and R. S. Kenett, "Joseph M. Juran, a perspective on past contributions and future impact," *Journal of Reliability Engineering International* 23 (2007), pp. 653–63.

30 R. Koch, *80/20 Principle: The Secret of Achieving More with Less* (New York: Currency, 1998); R. Koch, *Beyond the 80/20 Principle: The Science of Success from Game Theory to the Tipping Point* (London: Nicholas Brealey Publishing, 2020).

31 B. Sharp et al., *Marketing's 60/20 Pareto Law* (SSRN Elsevier, 2019), https://openresearch.lsbu.ac.uk/item/88vw1.

32 Again, this is just a small sample of this category: L. A. Adamic and B. A. Huberman, "Power-law distribution of the World Wide Web," *Science* 287 (2000), p. 2115; L. A. Adamic and B. A. Huberman, "Zipf's law and the Internet," *Glottometrics* 3 (2002), pp. 143–50; H. Ebel et al., "Scale-free topology of e-mail networks," (2002), http://arxiv.org/pdf/cond-mat/0201476.pdf.

33 A. Clauset et al., "On the frequency of severe terrorist events," *Journal of Conflict Resolution* 51 (2007), pp. 58–87; B. J. McGill et al., "Species abundance distributions: Moving beyond single prediction theories to integration within an ecological framework," *Ecology Letters* 10 (2007), pp. 995–1015.

34 R. Perline, "Strong, weak and false inverse power laws," *Statistical Science* 20 (2005), pp. 68–88.

35 T. Fluschnik et al., "The size distribution, scaling properties and spatial organization of urban clusters: A global and regional percolation perspective," *International Journal of Geo-Information* 5 (2016), p. 110.

36 A. Clauset et al., "Power-law distributions in empirical data," *SIAM Review* 51 (2009), pp. 661–703; L. Benguigui and M. Marinov, "A classification of the natural and social distributions; Part one: The descriptions," (2015), https://arxiv.org/abs/1507.03408; A. Corral and

A. Gonzalez, "Power law size distributions in geosciences revisited," *Earth and Space Science* 6 (2019), pp. 673–97.

37 M. Buchanan, "Laws, power laws and statistics," *Nature Physics* 4 (2008), p. 339.

38 J. M. Legare, "Achondroplasia," *GeneReviews* (October 1998; updated 2022), https://www.ncbi.nlm.nih.gov/books/NBK1152/pdf/Bookshelf _NBK1152.pdf.

39 Crédit Suisse, *Global Wealth Report 2021* (Zurich: Credit Suisse, 2021); Wealth-X, *Global HNW Analysis: The High Net Worth Handbook* (2019), https://www.wealthx.com/report/high-net-worth-handbook-2019/; Knight Frank, *The Wealth Report 2021*.

40 INSEE, "Niveau de vie moyen par décile," https://www.insee.fr/fr/ statistiques/2417897 [accessed 2021]; E. A. Shrider et al., *Income and Poverty in the United States: 2020* (Washington, DC: US Census Bureau, 2021).

41 A. H. Kent and L. Ricketts, "Has wealth inequality in America changed over time? Here are key statistics," Federal Reserve Bank of St. Louis, December 2, 2020, https://www.stlouisfed.org/open-vault/2020/december/has-wealth-inequality-changed-over-time-key-statistics.

42 V. Pareto, *Cours d'Économie Politique Professé a l'Université de Lausanne*, vol. II (Lausanne: Rouge, 1896), p. 312.

43 J. Han et al., "China's income inequality in the global context," *Perspectives in Science* 7 (2016), pp. 24–9; CEIC, "China's Gini Coefficient," 2021; S. Zhou and A. Hu, *China: Surpassing the "Middle Income Trap"* (Palgrave Macmillan, 2021).

44 E. A. Shrider et al., *Income and Poverty in the United States: 2020* (Washington, DC: US Census Bureau, 2021).

45 A. Walder, *China Under Mao: A Revolution Derailed* (Cambridge, MA: Harvard University Press, 2017).

46 K. Dolan, "Forbes' 35th Annual World's Billionaires List: Facts and figures," *Forbes*, April 6, 2021, https://www.forbes.com/sites/kerry adolan/2021/04/06/forbes-35th-annual-worlds-billionaires-list-facts-and-figures-2021.

47 G. Anthoons, *Iron Age Chariot Burials in Britain and the Near Continent* (Oxford: British Archaeological Reports, 2021).

48 Y. Zheng et al. (eds), *China's Great Urbanization* (London: Routledge, 2018).

49 United Nations, *World Cities Report 2020: The Value of Sustainable Urbanization*, https://unhabitat.org/sites/default/files/2020/10/wcr _2020_report.pdf.

50 B. Kiernan, *The Pol Pot Regime: Race, Power, and Genocide in Cambodia under the Khmer Rouge, 1975–79* (New Haven, CT: Yale University Press, 2008).

51 The most meaningful recent action was the reform of international taxation rules to ensure that multinational enterprises would pay a "fair" share of tax no matter where they operate.

52 US Geological Survey, "Comparison of materials emitted by notable volcanic eruptions," https://www.usgs.gov/media/images/comparison-materials-emitted-notable-volcanic-eruptions [accessed 2021].

53 V. Smil, *Global Catastrophes and Trends* (Cambridge, MA: MIT Press, 2008), p. 46.

54 CDC, "Past pandemics," https://www.cdc.gov/flu/pandemic-resources/basics/past-pandemics.html [accessed 2021].

55 Peculiarities of the COVID-19 pandemic included surprisingly large national and regional differences in infection rates and mortality, and major differences in the number of waves and their timing even among neighboring countries.

56 NASA, "Coronal mass ejections," https://www.swpc.noaa.gov/phenomena/coronal-mass-ejections [accessed 2021].

Acknowledgments

Besides (obviously) hundreds of scientists and artists whose cited writings and reproduced works made this book possible, my thank-you circle is small: Eva and David, the quartet at Penguin Random House (Connor Brown, Greg Clowes, Gemma Wain, Natalie Wall), and Bill Gates, for the past 14 years the most steadfast reader and reviewer of my books.

List of Illustrations

Index